The Autonomic Nervous System and Exercise

J. Hilary Green

Lecturer
Department of Human Movement and Recreation Studies
University of Western Australia

London
Chapman and Hall

First published in 1990 by
Chapman and Hall Ltd
11 New Fetter Lane, London EC4P 4EE

Typeset in 11/13pt Times by
EJS Chemical Composition,
Midsomer Norton, Bath

Printed in Great Britain by
T.J. Press Ltd, Padstow, Cornwall

ISBN 0 412 32500 4

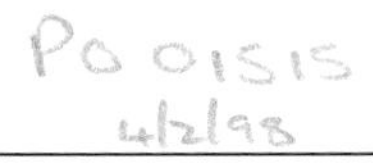

British Library Cataloguing in Publication Data

Green, J. Hilary
The autonomic nervous system and exercise.
1. Man. Exercise. Physiological aspects.
I. Title
612′044

ISBN 0 412 32500 4

To my parents

Contents

Preface ix

1 Organization of the nervous system 1
1.1 The neurone 1
1.2 Synapses 3
1.3 The action potential 7
1.4 Organization of the nervous system 10
1.5 Central nervous system 12
1.6 Peripheral nervous system 25
References and further reading 30

2 Physiology of the autonomic nervous system 32
2.1 Neurotransmission 32
2.2 Noradrenaline 34
2.3 Adrenal medulla 42
2.4 Assessment of sympathoadrenal activity 42
2.5 Acetylcholine 49
2.6 Non-noradrenergic non-cholinergic neural transmitters 53
References and further reading 53

3 Neuromuscular function 56
3.1 Skeletal muscle 56
3.2 Control of movement: peripheral mechanisms 62
3.3 Control of movement: central mechanisms 66
3.4 Autonomic function during exercise 68
References and further reading 71

4 Energy metabolism 72
4.1 Energy for muscular contraction 72
4.2 Metabolic pathways 79
4.3 Anaerobic provision of ATP for muscular work 91
4.4 Aerobic provision for ATP for muscular work 94
4.5 The autonomic nervous system and metabolism 97
References and further reading 101

5 The cardiovascular and respiratory systems 104

A Cardiovascular responses to exercise 104
5.1 The heart 104
5.2 Blood flow 113
5.3 Origin of the stimulus for autonomic nervous activity affecting the cardiovascular responses to exercise 116
5.4 Effect of autonomic nervous activity on the cardiovascular responses to exercise 118
5.5 Recovery from exercise 125

B Respiration during exercise 126
5.6 Pulmonary ventilation 126
5.7 Tissue respiration 131
5.8 Oxygen consumption 134
References and further reading 139

6 Temperature regulation 142
6.1 Body temperature 142
6.2 Heat balance 143
6.3 Measurement of body temperature 150
6.4 Physiological control mechanisms 153
6.5 Autonomic nervous control of thermoregulatory mechanisms and exercise 156
6.6 Models of thermoregulation 159
References and further reading 162

7 Factors affecting autonomic nervous activity 166
7.1 Ageing 166
7.2 Sex 169
7.3 Training 170
7.4 Drugs 172
References and further reading 174

8 Exercise and disease 178
8.1 Diabetes mellitus 178
8.2 Obesity 184
8.3 Atherosclerosis and ischaemic heart disease 187
References and further reading 195

Index 198

Preface

My main reason for writing a book on the role of the autonomic nervous system in exercise is to present an integrative approach to exercise physiology. Traditional textbooks on exercise physiology use what might be described as a systems approach. The inter-relationships between the systems, and therefore chapters, is often very difficult for the undergraduate student to grasp. This is difficult enough for a student of physiology. The problem is exacerbated for the sports science student who, in addition to studying applied physiology, is faced with other major disciplines as diverse as biomechanics and sociology. The present approach requires the reader to link areas of physiology in order to see how the major systems are all affected by autonomic neural control during exercise. The autonomic nervous system has been used as the link between the physiological systems challenged by exercise, not only because of its importance, but also because this seems to be an area which is given relatively little attention in traditional textbooks of exercise physiology.

The aim of Chapter 1 is to put the nervous system as a whole into perspective. Chapter 2 then focuses on the autonomic nervous system and considers its arrangement and physiology in detail. The third chapter discusses the role of the neuromuscular system in human movement, drawing attention to the roles of the central, peripheral afferent, efferent motor and autonomic nervous systems. The next three chapters go on to examine three areas in detail: energy metabolism, the cardiovascular and respiratory systems, and temperature regulation. These areas have been selected for their significance during exercise and because of the crucial importance of the autonomic nervous system in their control. In each case resting physiology is described before the derangements caused by exercise are discussed. The seventh chapter examines some of the factors which affect autonomic nervous activity during exercise, namely age, sex, training and drugs. The final chapter considers the clinical application of some of the applied physiology in the preceding chapters. The written text of each chapter is supported by tables and diagrams to elucidate the salient features.

The book is intended primarily for undergraduate and post-graduate students in sports science with a specialist interest in exercise physiology. It is assumed that the reader will already have studied some physiology and physiology of exercise, and therefore this text is intended to supplement general textbooks and lecture material. Since exercise physiology provides a good example of a disturbance of homeostasis and the subsequent role of physiological control mechanisms to restore equilibrium, it is hoped that this text will also prove useful for students in medical and life sciences.

I would like to thank Dr I. A. Macdonald, Reader in Physiology at the University of Nottingham, for his critical review of the first draft of this manuscript.

I would also like to thank Martin Thomson, Department of Anatomy and Human Biology, University of Western Australia, for drawing the original diagrams.

Chapter 1

Organization of the nervous system

1.1 The neurone

A nerve trunk comprises many single **neurones** (nerve cells). Thus, the nerve cell is the basic constituent of the nervous system (figure 1.1) and there are some 1×10^{12} neurones in the human nervous system. The neurone comprises three distinct portions, the **cell body**, the **axon** and the **nerve terminal**.

1.1.1 THE CELL BODY

The cell body of neurones ranges from 5–135 μm in diameter. In it are located the **nucleus**, **mitochondria**, **golgi apparatus** and **endoplasmic reticulum**. The nucleus consists of **deoxyribonucleic acid (DNA)** and **ribonucleic acid (RNA)** which control cell division and protein synthesis. The mitochondria have a vital role in the cell's oxidative metabolism. The Golgi apparatus has a secretory function and is continuous with the endoplasmic reticulum, a network of tubules. There are two kinds of endoplasmic reticulum, rough and smooth. The rough endoplasmic reticulum contains granules composed of RNA which are involved in protein synthesis. The smooth endoplasmic reticulum is concerned with steroid synthesis in the gonads and with detoxification in liver cells. In skeletal muscle and cardiac cells, the smooth endoplasmic reticulum forms the sarcoplasmic reticulum which plays a key role in muscle contraction (see section 3.1.3).

Arising from the cell body are projections called **dendrites**. Dendrites provide the receptive area of the cell to the nerve endings

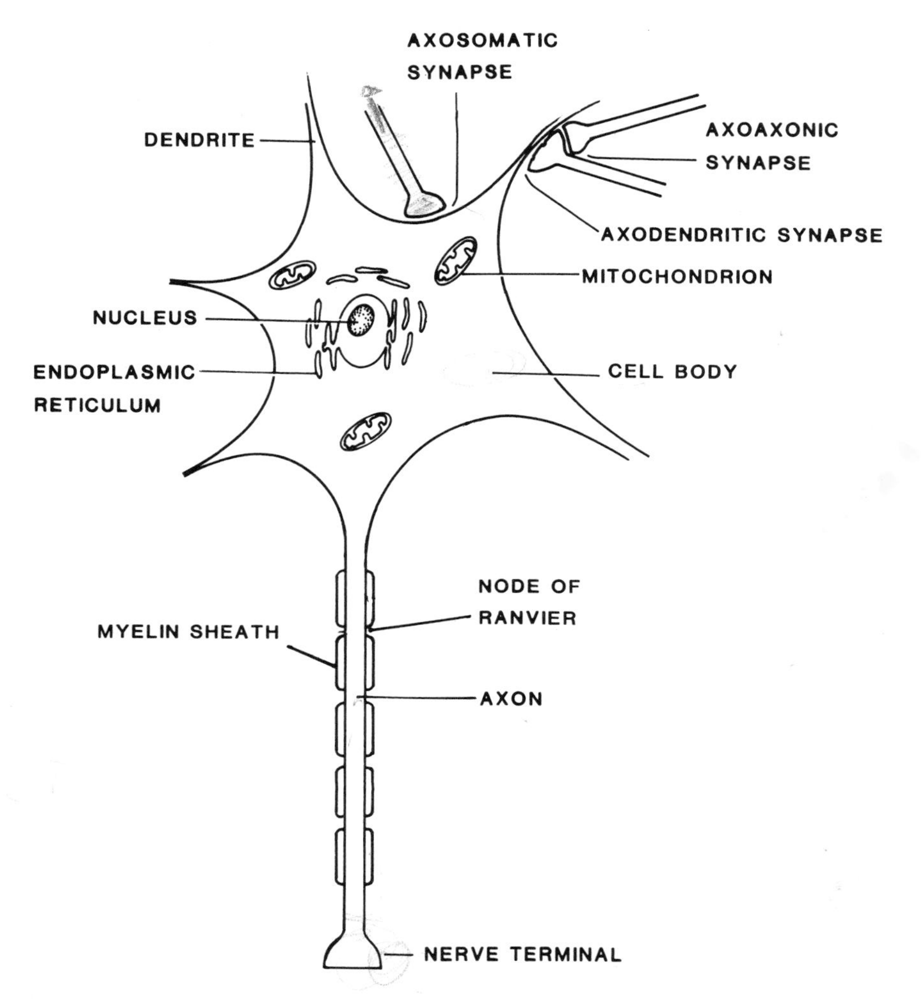

Figure 1.1 The neurone and synapse.

of other cells. The more surface area provided by the dendrites the greater the potential for incoming neural signals. Over three-quarters of the surface area of a spinal motor neurone may be due to dendrites. A spinal motor neurone is a neurone which leaves the spinal cord to end on a skeletal muscle fibre. Firing of the spinal motor neurone is the trigger for muscle contraction. The cell bodies of nerve cells give rise to the grey matter of the brain and spinal cord.

1.1.2 THE AXON

The axon is a tube of cytoplasm within a thin protein and lipid membrane through which protein formed in the cell body is transported to the nerve terminal. This may occur through the longitudinally running microtubules and neurofilaments in the axon, although the function of these tubular structures is not completely understood. The proteins that are transported down the axon include enzymes and vesicle membrane material which have a vital role in neurotransmission (section 2.1). The axon plays a crucial role in the propagation of the nerve impulse. Large axons (1–25 μm in diameter) are covered with a myelin sheath which is broken at intervals by **nodes of Ranvier** where the axon is exposed. The nodes of Ranvier are less than 1 μm wide. The internodal sections vary in length (200–1500 μm) in relation to the length of the axon, with longer axons having longer internodal sections. In the central nervous system the myelin sheath is made of glial cells called **oligodendrocytes** which are white and account for the white matter of the brain and spinal cord. In the peripheral nervous system the myelin sheath is made of **Schwann cells**. The myelin sheath provides a layer of electrical insulation for the cell. Its thickness is related to the size of the axon with larger axons having thicker myelin sheaths. The small axons (< 1μm in diameter) have no sheath at all.

1.1.3 THE NERVE ENDING

The nerve ending is a specialized structure which communicates with a second neurone or tissue in a **synapse**.

1.2 Synapses

Communication between nerve cells and between a nerve cell and a target cell is by chemical and/or electrical synapses. Synapses between neurones may be **excitatory** or **inhibitory**.

A synapse occurs when the end of one axon makes chemical or electrical contact with the cell body (**axosomatic synapse**), dendrite (**axodendritic synapse**) or nerve terminal (**axoaxonic synapse**) of another neurone. In some cases a synapse may occur between two dendrites (**dendrodendritic synapse**) but the axosomatic and axodendritic synapses are the more common. The first neurone is termed **presynaptic** and the second **postsynaptic**.

Some neurone connections involve three neurones. Here the synapse between the first and second neurone may be axoaxonic and the synapse between the second and third neurone may be axodendritic (figure 1.1). Other synapses involve reciprocal activity where a neurone can act as both a pre- and postsynaptic neurone.

In chemical communication, the more common method, **neurotransmitters** are released from the nerve ending in response to electrical activity along the axon (figure 1.2a). The neurotransmitter is a molecule for which receptors on the target cell membrane have a specific affinity. Neurotransmitter molecules are stored in vesicles in the nerve terminal and they are released on the arrival of a nerve impulse. The vesicles of some nerve terminals also contain substances which modulate synaptic activity and which are called **neuromodulators**. Humoral substances (including many hormones) also act as neuromodulators. The neurotransmitter molecules cross the **synaptic cleft** (the tiny gap between pre- and postsynaptic neurones) to bind to the receptor on the postsynaptic membrane causing a change in the cell membrane permeability to inorganic ions. There is always a brief delay, called **synaptic delay**, between the arrival of the presynaptic impulse and the change in postsynaptic membrane permeability due to the emptying of the neurotransmitter molecules from the storage vesicles. Finally the neurotransmitter is inactivated by a specific enzyme and/or is taken back up by the presynaptic nerve ending.

The central neurotransmitters include **noradrenaline**, **dopamine**, γ **aminobutyric acid (GABA)**, **glycine**, **5-hydroxytryptamine (5-HT)**, **glutamate** and **aspartate**. The principal neurotransmitters in the peripheral nervous system are **noradrenaline** and **acetylcholine**. These are discussed in Chapter 2.

The physical and chemical properties of neurotransmitter receptors are not fully understood. However, it is thought that the neurotransmitter fits the receptor spatially rather like a lock and key, and that it may also be attracted to the receptor by electrostatic forces. Thus the neurotransmitter and the receptor form a complex.

Normally electrical activity travels only in one direction along a nerve axon towards the cell body of the postsynaptic neurone (although, experimentally, it is possible to electrically stimulate the middle of an axon and cause impulses to travel in both directions away from the point of stimulation). When a nerve impulse reaches a chemical synapse, it is transmitted in one direction only, along the

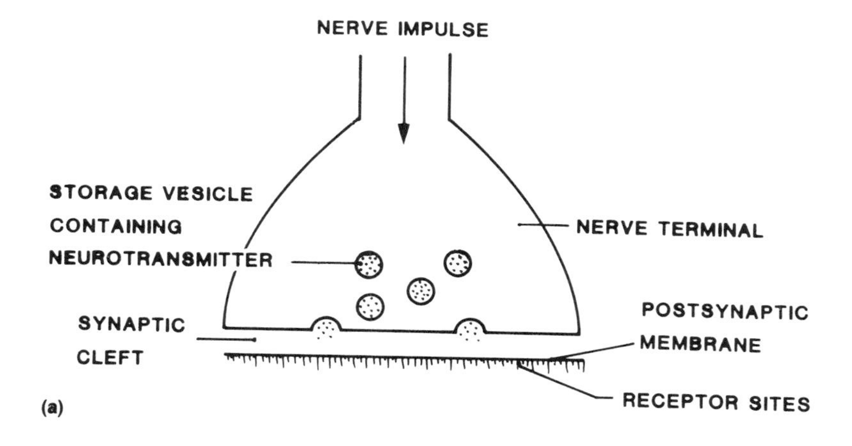

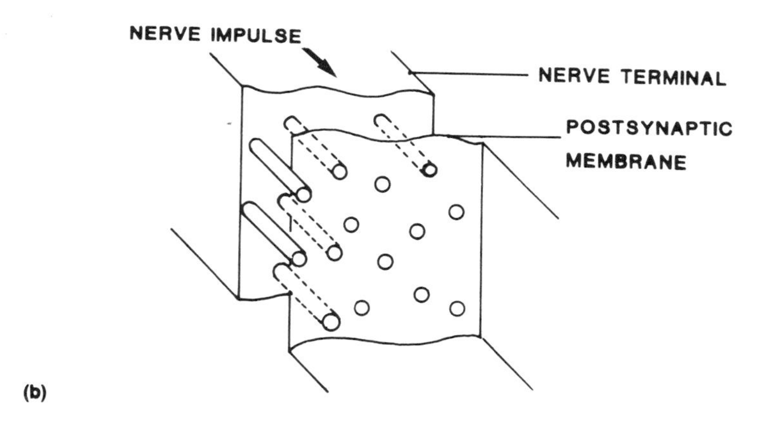

Figure 1.2 Schematic arrangement of neurotransmission. (a) Chemical synapse; (b) electrical synapse.

postsynaptic neurone. In contrast transmission occurs in either direction between two cells linked by an electrical synapse (found, for example, between some neurones of the sensory cortex, olfactory bulb and retina). In an electrical synapse the two neurones lie much closer together than in a chemical synapse. They are linked by channels in the cell membrane which allow the direct passage of ions and small molecules between the cells. When two neurones join in this way they form **gap junctions** (figure 1.2b). Some synapses involve both chemical and electrical transmission.

1.2.1 IONIC EVENTS AT A SYNAPSE

Ion distribution in the intra- and extracellular fluid

The cell membrane is formed from a double layer of phospholipid molecules in which there are hydrophilic channels that allow the passage of ions.

The interior of the nerve cell (like all cells) is negatively charged relative to the outside as a result of the ease with which K^+ can cross the cell membrane and of the intra- and extracellular concentrations of ions. The interior of the cell is rich in K^+ but low in Cl^- and Na^+ compared with the outside.

In this state the cell is polarized. That is, there is a potential difference across the plasma membrane. The resting potential of a nerve cell is in the order of -70 mV inside relative to outside.

Excitatory synapse

In an excitatory synapse the cell membrane potential is moved closer to the threshold potential due to an increased membrane permeability for Na^+. When the excitatory synapses predominate, they initiate the nerve impulse in the postsynaptic membrane (figure1.3a). The nerve impulse then travels down the axon (section 1.3.1.).

Inhibitory synapse

In an inhibitory synapse the postsynaptic cell membrane permeability for Na^+ is unchanged. Instead the action of the neurotransmitter is to increase the membrane permeability for K^+ and/or Cl^-, which causes the cell membrane potential to become more negative and therefore further away from the threshold level for excitation (figure 1.3b). The only known inhibitory neurotransmitters are GABA (found in the cerebral cortex and cerebellum) and glycine (found in the spinal cord).

1.2.2 NEURAL INTEGRATION

It is important to recognize that a single excitatory synapse is insufficient to cause the postsynaptic neurone to fire. Hundreds of **excitatory postsynaptic potentials (EPSPs)** are required to achieve

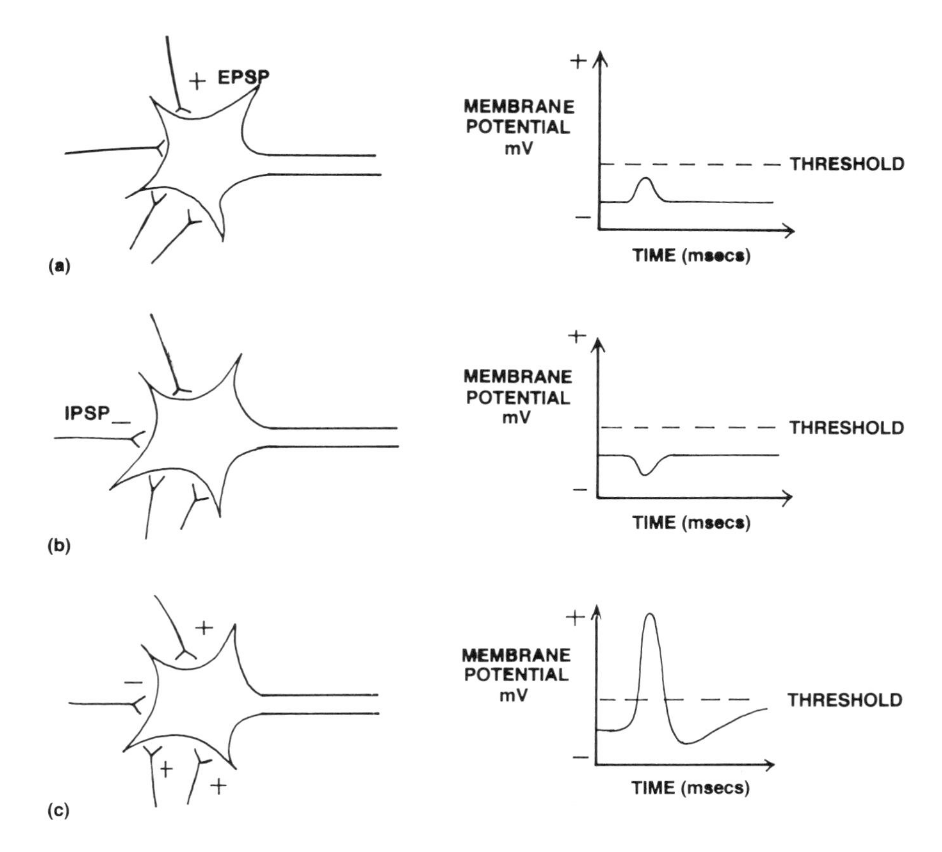

Figure 1.3 Neural integration. (a) Excitatory postsynaptic potential (EPSP); (b) inhibitory postsynaptic potential (IPSP); (c) action potential: dominance of EPSPs sufficient to raise the membrane potential above threshold.

this. The net effect of EPSPs and **inhibitory postsynaptic potentials (IPSPs)** determines whether the wave of excitation will continue along the postsynaptic neurone (figure 1.3c).

1.3 The action potential

As mentioned previously, when nerves are stimulated, the cell membrane potential becomes less negative and the membrane becomes more permeable to Na^+ ions which enter the cell via specific channels in the cell membrane, under the influence of Ca^{2+}. This causes the interior of the cell to become more positively charged relative to the outside. The cells are thereby depolarized. This reversal of net charge across the membrane is called an **action potential**. When the Na^+ channels are open, the cell is absolutely refractory (it will not conduct a second impulse). Within milliseconds the permeability to Na^+ is reversed, that is the Na^+ channels close, so that no further entry of Na^+ is possible. In this state the cell is relatively refractory (it will conduct a second impulse only if the strength of the stimulus is increased). Less than a millisecond later the cell's permeability to K^+ increases, which results in the loss of intracellular K^+ via specific K^+ channels. This restores the baseline electrical potential. However, this electrical potential is actually overcompensated by a transient accumulation of extracellular K^+ leading to a more negative after-potential relative to the baseline state. The ionic balance is subsequently restored by an ATP-requiring Na^+ pump, an Na^+–K^+ ATPase in the cell membrane. The ATPase is a lipoprotein, and ATP is generated from oxidative carbohydrate metabolism. Further consideration of ATP is given in section 3.1.3. Since this pump is more effective for Na^+ than for K^+, complete restoration of the ionic balance also requires diffusion of K^+ across the nerve membrane, which becomes more permeable to K^+ (figure 1.4).

1.3.1 PROPAGATION OF THE NERVE IMPULSE

A nerve impulse is transported along the axon with the current of electricity at right angles to the direction of movement of the impulse. Although the ion movements during the impulse are passive, they are enabled by metabolic activity and therefore should not be thought of as a process of passive conductance. In myelinated fibres the nerve impulse jumps between the nodes of Ranvier by **saltatory conduction** (*saltare* (Latin) means 'to jump'). In non-myelinated fibres, depolarization of one part of the cell leads to depolarization of adjacent parts. This is known as **cable conduction**. Conduction in non-myelinated fibres is much slower than in myelinated fibres. The distance travelled along an axon depends on the individual cell, with some axons extending for as long as 1 m from the spinal cord to a peripheral muscle in the case of spinal motorneurones.

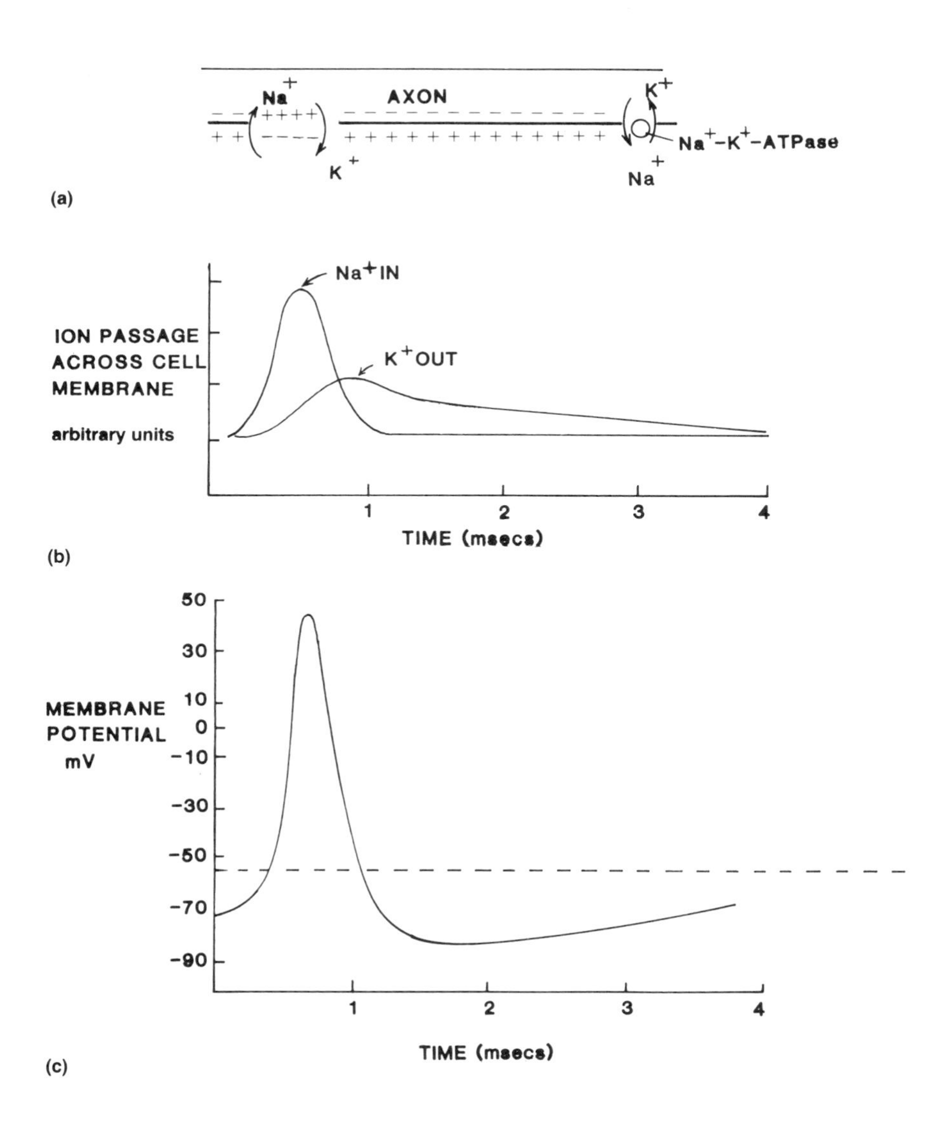

Figure 1.4 The action potential. (a) Change in membrane potential due to passage of ions across the cell membrane; (b) conductance of ions across the cell membrane with time during an action potential; (c) the action potential.

1.4 Organization of the nervous system

The **central nervous system** includes the brain and spinal cord. The **peripheral nervous system** includes all nerves outside the brain and spinal cord, namely the cranial and spinal nerves. These are mixed nerves, containing both neurones which travel towards the brain (**afferent** neurones) and those which travel away from the brain (**efferent** neurones). The afferent neurones relay sensory information from the organ in which their sensory ending is located. The efferent neurones are of two main types, **somatic** and **autonomic** (table 1.1).

Table 1.1 Organization of the nervous system

A. Central nervous system
   1. Brain
   2. Spinal cord

B. Peripheral nervous system
   1. Afferent
   2. Efferent
      (a) Somatic
      (b) Autonomic
         (i) Sympathetic
         (ii) Parasympathetic

The somatic neurones innervate skeletal muscle (**voluntary**) while the autonomic neurones innervate smooth muscle such as the intestines and cardiac muscle (**involuntary**), as well as sweat and salivary glands (exocrine glands) and some endocrine glands. Firing of the somatic neurones is always excitatory and therefore brings about muscle contraction. Firing of the autonomic neurones may be excitatory or inhibitory. In general excitatory responses are mediated by **sympathetic fibres** and inhibitory responses are mediated by **parasympathetic fibres**. Notable exceptions are the vagal (parasympathetic) control of gastrointestinal motility and tone, and secretion of insulin from the pancreas, which are excitatory responses.

1.4.1 CLASSIFICATION OF NERVE FIBRES

Nerve fibres are classed as **A** (**α**, **β**, **γ** and **δ**), **B** or **C** according to their diameter (which determines their rate of conduction). Sensory fibres are additionally classed as **I**, **II**, **III** and **IV**. The myelinated fibres of the motor nerves are the fastest conducting fibres with conduction velocities of up to 120 m s^{-1}. The preganglionic autonomic fibres conduct at a velocity of around 9 m s^{-1} and the non-myelinated postganglionic fibres at around 1.4 m s^{-1} (table 1.2; see section 1.6.3 for a definition of pre- and postganglionic fibres).

The autonomic nervous system will be described in more detail in Chapter 2. The present chapter concentrates on the organization of the nervous system as a whole.

Table 1.2 Classification of nerve fibres (m = myelinated, um = unmyelinated, GTO* = Golgi tendon organ)

Fibre	Class	Function	Diameter (μm)	Conduction velocity (m s^{-1})
Aα (m)	I	Afferent fibres for GTO and some muscle spindles		
	—	Efferent somatic motor fibres	12–20	70–120
Aβ (m)	II	Afferent fibres for touch pressure and some muscle spindles	5–12	30–70
Aγ (m)	—	Efferent motor fibres to muscle spindle	3–6	15–30
Aδ (m)	III	Afferent fibres: pain, temperature and touch	1–5	5–15
B (m)	—	Efferent preganglionic autonomic fibres	1–3	3–15
C (um)	IV	Afferent fibres for pain and temperature		
	—	Efferent postganglionic autonomic fibres	0.5–1.5	0.6–2.2

* See section 1.5.3.

1.5 Central nervous system

1.5.1 BRAIN

The brain is divided into six main areas:

medulla oblongata
pons
midbrain
cerebellum
diencephalon
telencephalon

The medulla, pons and midbrain constitute the **brainstem** (figure 1.5).

1.5.2 THE BRAINSTEM (MEDULLA, PONS AND MIDBRAIN)

Medulla

The medulla (medulla oblongata) is situated immediately next to the spinal cord and extends approximately 2 cm to the pons. The ventral surface of the medulla forms the pyramid through which the corticospinal tract runs (section 1.5.3). Four of the 12 cranial nerves have their origin in the medulla. These are:

IX, **glossopharyngeal**
X, **vagus**
XI, **accessory**
XII, **hypoglossal**

Pons

The pons (bridge between the two hemispheres of the cerebellum) extends about 2.5 cm beyond the medulla. The pons is divided into two sections; the dorsal and ventral portions. Four cranial nerves have their origin in the dorsal portion of the pons and these are:

V, **trigeminal**
(ophthalmic, maxillary, mandibular)
VI, **abducent**
VII, **facial**
VIII, **vestibulocochlear**

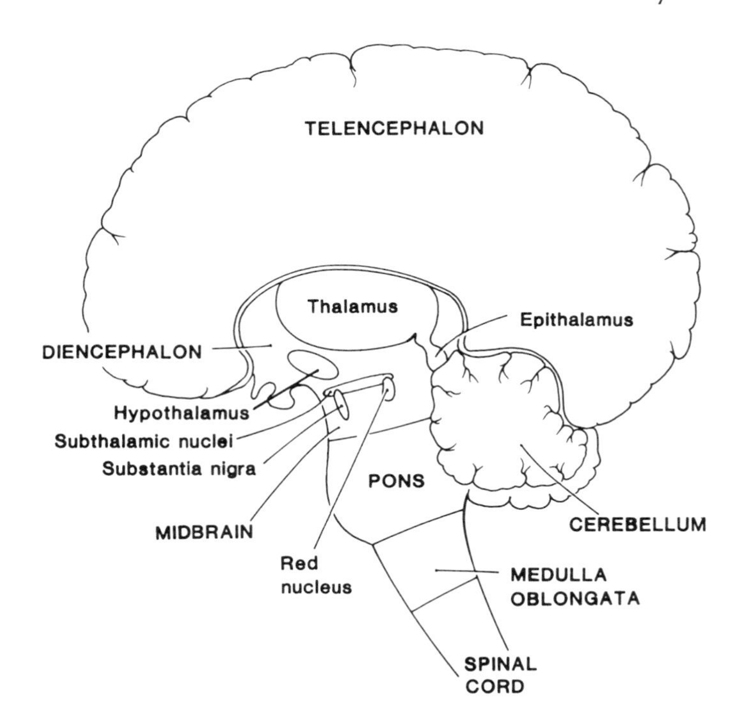

Figure 1.5 Principal divisions of the brain.

The ventral (basal) portion forms a neuronal junction connecting nerve fibres between the cerebral cortex and cerebellum.

Midbrain

The midbrain is about 1.5 cm long and is connected to the cerebellum. It forms a link between the pons and the cerebral hemispheres. Two of the four remaining cranial nerves originate in the midbrain. These are:

III, **oculomotor**
IV, **trochlear**

Also situated in the midbrain are the **red nucleus** and **substantia nigra**, both motor nuclei forming part of the **extrapyramidal motor system** (section 1.5.3).

Reticular formation The **reticular formation** is a network of afferent and efferent neurones in the core of the brainstem. It contributes to sleep/arousal and to motor and visceral functions. It extends beyond the brainstem into the **hypothalamus** and part of the **thalamus**.

Hindbrain The medulla and cerebellum together constitute the **hindbrain**.

Cerebellum

The cerebellum is about 10 cm wide, 5 cm long and 4 cm thick. It comprises two lateral **hemispheres** and the **vermis** in the centre. The cerebellum receives afferent input but its major role is in motor function. In this context its neurones play a key role in the control of balance and coordination.

Diencephalon

Immediately above the midbrain is the diencephalon which forms a core within the cerebral hemispheres. The main structures of the diencephalon are:

thalamus
hypothalamus
epithalamus
subthalamus

Thalamus The two thalami are each approximately 3 cm long, 1 cm wide and 1 cm high. The thalamus influences autonomic function through its anatomical connections with the hypothalamus. There are further connections between the thalamus and cerebellum and the thalamus and cerebral cortex. Through these connections the thalamus acts as a relay station between the cerebellum and cerebral cortex. It thus plays a key role in sensorimotor function.

Hypothalamus The hypothalamus is situated below the thalamus. Although weighing about only 4 g, the hypothalamus plays a critical role in regulating key body functions including metabolism and body temperature. It also influences the activity of the autonomic nervous system. Hypothalamic activity is influenced by neuronal input from the thalamus and limbic brain system, a group of structures situated around the brainstem (*limbus* means a ring), as well as by circulating blood temperature and hormone levels. Hypothalamic activity is also modulated by changes in arterial blood pressure and gas tensions via receptors located in the aorta and coronary arteries.

Epithalamus The **epithalamus** comprises the **habenula nuclei** and the **pineal gland**. The habenula nuclei are involved in the visceral effects of emotional drives and smell. The importance of the pineal gland is still poorly understood.

Subthalamus The motor nuclei, substantia nigra and red nucleus extend from the midbrain into the subthalamus. The subthalamus also includes the **subthalamic nucleus**, which is a motor nucleus of the extrapyramidal system.

Telencephalon

The telencephalon is the area of the brain occupied by the two hemispheres of the **cerebral cortex**, the **corpus striatum** and **medullary centre**. The cerebral cortex constitutes about 40% of the total brain weight. The cells of the cerebral cortex provide the brain's shell, which is about 3 mm deep. The cortical cells are arranged in layers. There are three main types of nerve connection in the cortex, which are situated in the medullary centre, the white matter beneath the grey cortex. These are:

projection neurones which relay nerve impulses to subcortical centres
association neurones which relay nerve impulses within the cerebral cortex in the same hemisphere
commissural neurones which relay nerve impulses within the cerebral cortex but to the opposite hemisphere

The cerebral cortex is divided into four lobes, the **frontal**, **temporal**, **parietal** and **occipital** lobes (figure 1.6).

The cortical cells have specialized sensory and motor functions as well as functions involving behaviour. The cells that are involved in these various functions are localized in areas now known as the sensory, motor and association cortex.

Beneath each cerebral hemisphere, and in close association with the thalamus, is the corpus striatum. The corpus striatum includes the

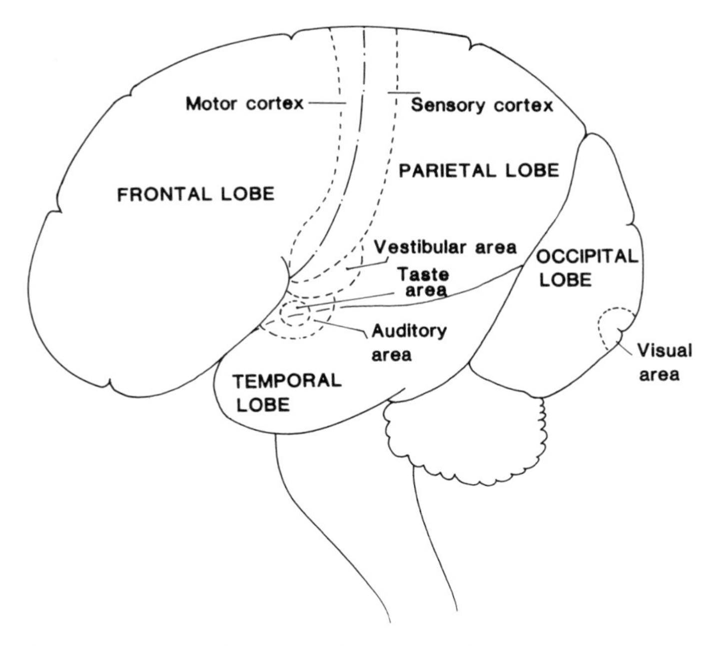

Figure 1.6 Four lobes of the cerebral cortex and sensory areas.

caudate nucleus and **lentiform nucleus** (figure 1.7). The corpus striatum and the motor nuclei, subthalamic nucleus and substantia nigra constitute the basal ganglia which play an important role in motor control.

Limbic system The **limbic system** comprises:

frontal lobe of cerebral cortex
temporal lobe of cerebral cortex
thalamus
hypothalamus and their neural connections

This system of structures is involved with emotional behaviour and learning.

Ventricles The **ventricles** are cavities in the brain which are continuous with the central canal of the spinal cord. Each of the four ventricles contains **cerebrospinal fluid** produced from the **choroid plexus** of each ventricle. The choroid plexus is located in the roof of

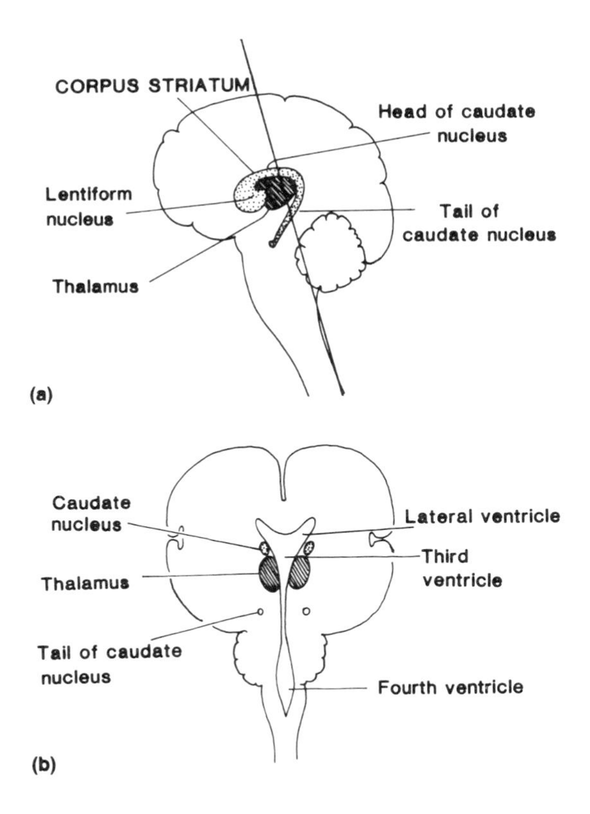

Figure 1.7 Location of the corpus striatum (schematic).

each ventricle and, as well as secreting cerebrospinal fluid, is the means by which matter is exchanged between the blood and cerebrospinal fluid.

1.5.3 THE SPINAL CORD

The spinal cord is about 45 cm long and about 1 cm in diameter and is made up of 32 **vertebrae** (seven **cervical**, twelve **thoracic**, five **lumbar**, five **sacral** and four **coccygeal**). When viewed in cross-section the spinal cord has an H-shaped core of grey matter. The limbs of this H-shaped core are known as the **ventral** (anterior) and **dorsal** (posterior) **horns** and contain principally three types of neurone: interneurones, Aα motor neurones and Aγ motor neurones. The motor neurones of the ventral horn supply **extrafusal** (Aα) and **intrafusal** (Aγ) skeletal muscle fibres. Sensory (afferent) neurones enter the spinal cord by way of the dorsal horn. An intermediary lateral horn (situated between the ventral and dorsal horns) is found in the thoracic and upper three lumbar segments. The preganglionic fibres of the sympathetic autonomic neurones (see section 1.6.3) have their origin in the lateral horns of these segments. Surrounding the grey core is an area of white matter which contains the **ascending** and **descending nerve tracts**.

1.5.4 ASCENDING AND DESCENDING NERVE TRACTS

The nerve fibres of the central nervous system are arranged in bundles known as tracts or pathways (figure 1.8).

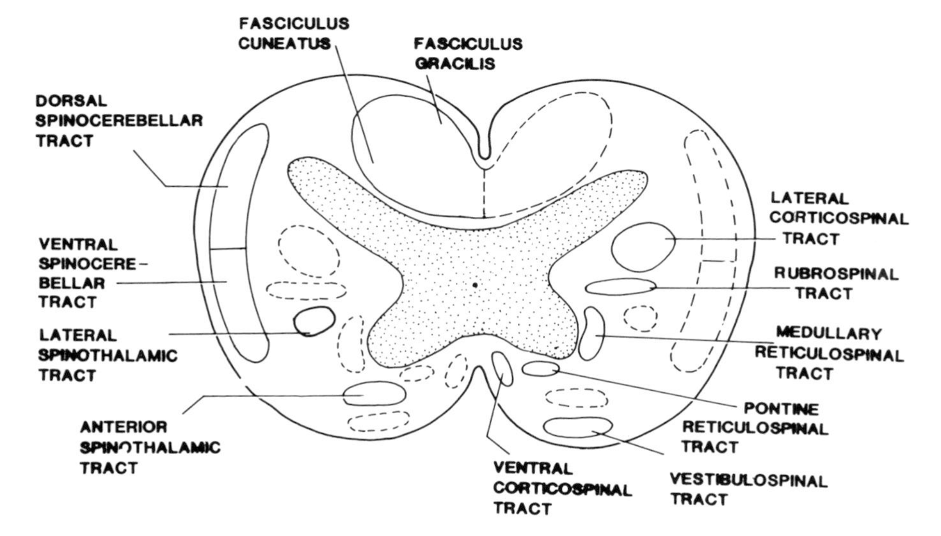

Figure 1.8 Transverse section of spinal cord showing the major ascending (left side) and descending (right side) nerve tracts at C4. Figure adapted from Barr and Kiernan (1983).

Ascending nerve tracts

Ascending nerve tracts in the spinal cord relay sensory information from peripheral sensory receptors to the central nervous system. Sensory pathways are typically comprised of three neurones. The first has its cell body in a dorsal root ganglion from which axons pass to the receptor and to the spinal cord. The cell body of the second neurone is in the spinal cord and its axon passes up the spinal cord to the thalamus where the third neurone has its cell body. The axon of the latter neurone passes up to the cerebral cortex (figure 1.9). Thus, the first neurone is part of the peripheral nervous system while the second two are in the central nervous system.

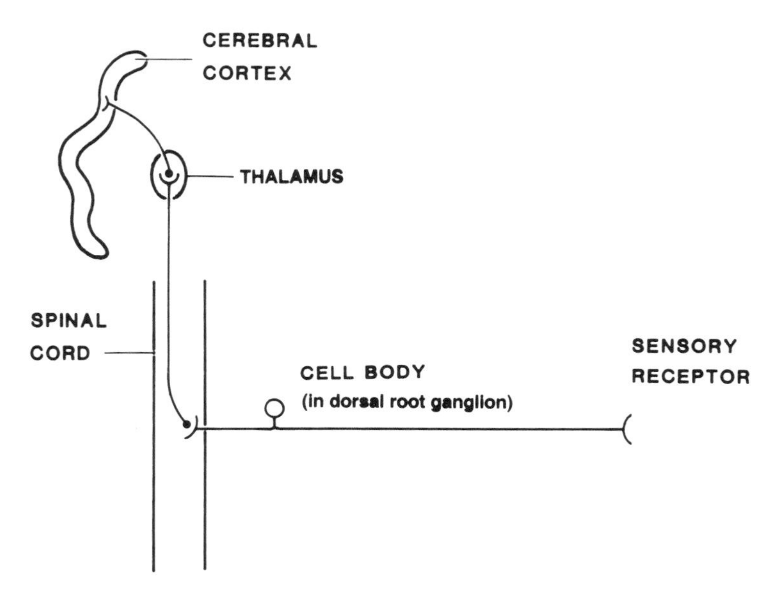

Figure 1.9 Typical three-neurone arrangement of the sensory nervous system.

The sensory receptors

The peripheral sensory nerve endings form specialized receptors to sensory information. They respond to both conscious and subconscious sensation. The receptors for conscious sensation are involved in position sense (**kinaesthesia** or **proprioception**), temperature and pain as well as Aristotle's original five senses: sight, sound, smell, taste and touch. In higher vertebrates, such as man, visual sensation is highly developed. However, in lower vertebrates, such as the dog with its less well-developed cortex and associative areas but abundance of olfactory receptors, the sensation of smell is more important. Receptors for unconscious sensation, that is changes in the body's internal environment, are the **chemoreceptors** which are sensitive to blood gas tensions (**pCO_2**, **pO_2**) and **pH** and the

baroreceptors which are sensitive to the pressure exerted by blood on the artery wall in which the receptors are located.

Position sense (kinaesthesia/proprioception) The receptors for position sense include the **muscle spindles** which respond to stretch and regulate stiffness, the **Golgi tendon organs** which respond to high muscle tension and stretch, the **joint receptors** which respond to joint movement, and the **vestibular apparatus** of the inner ear which responds to postural movement. The main ascending pathways for the axons of these receptors are located in the **dorsal column**. There are two pathways for proprioception. Pathways for conscious proprioception are located in the **fasciculus gracilis** and **fasciculus cuneatus** columns. The axons of neurones in this pathway end in the cerebral cortex in the typical manner. The pathways for unconscious proprioception are located in the dorsal and ventral spinocerebellar tracts. The axons in this case end in the cerebellum rather than in the cerebral cortex.

Temperature The peripheral thermoreceptors are located just below the skin. There are two kinds of peripheral temperature receptor, 'cold' and 'warm'. The 'cold' receptors fire in response to local temperatures between 10 and 35°C with a maximum frequency of discharge at around 30°C. Paradoxically these fibres also fire at temperatures above 45°C. For this reason rapid warming of the skin to around 45°C induces a sensation of coldness. This phenomenon is termed the **paradoxical cold response**. By contrast, the 'warm' receptors fire in response to local temperatures between 35 and 45°C, with a maximum frequency of discharge at around 40°C. The axons of the temperature receptors travel up the spinal cord in the lateral spinothalamic tract to the thalamus and thereafter make neural connections with the sensory cortex and hypothalamus where the central temperature receptors are located. The roles of the peripheral and central thermoreceptors will be discussed more fully in Chapter 6.

Pain The receptors for pain (**nociceptors**) are free nerve endings with varying degrees of density according to the sensitivity of the organ. The nociceptive fibres fire in response to various stimuli: mechanical (such as a physical knock), thermal (excessive heat or cold) or chemical. A chemical stimulus may be extrinsically applied (such as toxic fumes) or intrinsically generated (including 5-HT (serotonin), histamine, pancreatic kallikrein and trypsin, plasma kinins and prostaglandins). Neural signals originating in these nociceptors are transmitted by type Aδ (acute sharp pain) or type C (chronic burning pain) afferent fibres up the spinal cord in the spinothalamic tract with the temperature fibres to the thalamus. From the thalamus fibres ascend to widespread areas of the sensory cortex.

Sight The visual receptors are located in the **retina** of the eye. There are two kinds of receptor, **rod cells** and **cone cells**. The rod cells fire in low-intensity light and are not receptive to colour. In contrast, the cone cells respond both to light and to colour. Neural information from the rod and cone cells is carried in the **optic nerve** (cranial nerve II) to the **optic chiasm**. Here the fibres from each side cross and travel up the **optic tracts**, mostly to the **visual cortex** in the **occipital lobe**. Due to the crossing of fibres in the optic chiasm the right visual cortex receives neural signals from the left eye and vice versa.

Sound Cells that are receptive to sound are located in the **organ of Corti** in the **cochlea** of the inner ear and respond to vibration. They have no axons but form connections via peripheral processes with the afferent fibres of the vestibulocochlear nerve (cranial nerve VIII) which travels in the **auditory tract** to the **auditory cortex** of the **temporal lobe**.

Smell (olfaction) The olfactory receptors are located in the **nasal mucosa**. When stimulated they fire and the signal is relayed via the olfactory nerve (cranial nerve I) to three olfactory areas in the brain.

Taste (gustation) The taste receptors, or **taste buds**, are mainly located on the tongue. They are innervated by a branch of the facial nerve (cranial nerve VII) and by branches of the glossopharyngeal nerve (cranial nerve IX). They relay information into the nucleus of the **solitary tract** in the medulla oblongata. From the nucleus of the solitary tract, fibres run to the thalamus and on to the taste area in the **parietal cortex**.

Touch There are various receptors for tactile sensation in the skin. The afferent pathways are located in the dorsal columns (**fasciculus**

gracilis and **fasciculus cuneatus**), and in the anterior spinothalamic tract. The pathways end in thalamic nuclei from which projections travel to the sensory cortex.

Blood chemistry Receptors that detect changes in blood chemistry play a crucial role in the control of breathing by transmitting neural information to the medullary inspiratory neurones.

pO_2 The **carotid** and **aortic bodies** are sensitive to changes in pO_2. These receptors are located at the bifurcation of the common carotid artery and in the arch of the aorta. When pO_2 is low, both receptors fire but the carotid bodies have the more significant role. If an individual breathes normal room air (21% oxygen) after breathing 100% oxygen, ventilation increases by about 20%. Further decreases in the percentage of oxygen inhaled have no effect on ventilation until the inspiratory oxygen content has decreased to around 10%. Thus, it takes a large drop in inspired oxygen to cause an increase in ventilation. Such a low pO_2 does not occur even in strenuous exercise, and it is therefore not the stimulus for increased ventilation during exercise.

pCO_2 In contrast to the tolerance shown for low inspired oxygen, man displays a poor level of tolerance for elevated levels of inspired carbon dioxide. This poor tolerance of carbon dioxide seems to be the major factor which terminates breath holding. The medullary neurones do not detect the change in pCO_2 *per se* but rather respond to changes in pH. During exercise the changes in pCO_2 are small or not detectable, even when ventilation is increased several-fold. Thus an increase in mean arterial pCO_2 is not the main stimulus for increasing ventilation during exercise.

pH There are two groups of H^+ receptor, a peripheral group and a central group. The peripheral chemoreceptors are again the carotid bodies and they play an important role in mediating an increase in ventilation due to an increase in arterial H^+ ion concentration (i.e. decrease in arterial blood pH) in the absence of an increase in pCO_2. The central chemoreceptors are not involved since H^+ ions pass the blood–brain barrier poorly. By comparison with this H^+ reflex, the carotid bodies play a more important role in the pO_2 reflex in which a decrease in arterial pO_2 causes an increase in the firing of the carotid bodies. Through their neural input to the medullary respiratory centre the carotid bodies bring about an increase in ventilation. Patients who have had their carotid bodies removed fail to display the normal increase in ventilation during exercise, especially at high work loads. Thus, the carotid bodies are thought to play a role in the control of ventilation during exercise.

The likely site of the central chemoreceptors is the ventral surface of the medulla. The importance of these receptors seems to be in maintaining brain pH. For example, if alveolar pCO_2 is increased due to an increased pCO_2 of inspired air, the CO_2 diffuses across the blood–brain barrier leading to an increase in pCO_2 in the cerebrospinal fluid (CSF). The resultant decrease in CSF pH stimulates the central chemoreceptors which, via their neural input to the medullary respiratory centre, bring about an increase in ventilation. This increase in ventilation lowers arterial pCO_2 and subsequently, therefore, CSF pH increases. However, any response to a decrease in peripheral pH independently of an increase in pCO_2 is mediated by the peripheral chemoreceptors since H^+ ions penetrate the blood–brain barrier poorly. Thus, breathing is critically important for controlling brain pH. For a recent review of central chemoreceptor function see Eugene and Cherniak (1987).

Baroreceptors The baroreceptors are sensory nerve endings located in the walls of the large arteries. They are especially abundant in the aorta and carotid artery. They respond to stretching of the arterial wall due to pressure. Thus, when blood pressure (the pressure exerted on the arterial wall in which the receptors are located) increases, the baroreceptors are stimulated, and increased neural discharge from them is transmitted to the medulla. In the case of baroreceptors in the aorta, this is via sensory neurones in the **vagus nerve** (cranial nerve X). In the case of baroreceptors in the carotid artery this is via the **sinus nerve** (also called **Hering's nerve**) which is a branch of the glossopharyngeal nerve (cranial nerve IX). The increase in neural activity reaching the medulla via these nerves inhibits firing of the vasoconstrictor neurones, leading to passive vasodilatation and therefore a lowering of arterial blood pressure.

Descending nerve tracts: motor systems

Specific central nervous structures and peripheral nerves form circuits which bring about movement. The central nervous structures involved in motor control include the cerebral cortex, basal ganglia (corpus striatum, subthalamic nucleus and substantia nigra), thalamus, red nucleus, reticular formation and the lateral vestibular nucleus. Motor control requires sensory information from peripheral receptors. The motor outflow from these central structures reaches the muscle via one of two systems or pathways, the pyramidal system (direct corticospinal pathway); and the extrapyramidal system (multineuronal pathway). These descending pathways are principally controlled by nerve fibres from the cerebral cortex, cerebellum and basal ganglia, and these central structures are discussed in more detail in Chapter 3.

The pyramidal system The fibres of the pyramidal system originate in the cerebral cortex and pass through the medullary centre and the internal capsule which separates the lentiform nucleus from the thalamus. The fibres continue to descend through the brainstem, and when they reach the pyramid of the medulla about 85% of the fibres cross to pass down the opposite side, thus forming the lateral corticospinal tract. The remaining fibres continue to pass down the spinal cord on the same side, thus forming the ventral **corticospinal tract**. The transfer of fibres to the opposite side is known as **pyramidal decussation**. The fibres of these tracts influence motor neurones principally via interneurones, but also synapse directly with motor neurones. Historically, the corticospinal tracts were discovered before other motor tracts. It was therefore considered for many years to be *the* motor system. Other tracts with motor function were therefore termed extrapyramidal.

The extrapyramidal system In addition to nerve tracts arising from the cerebral cortex, further tracts arise from the red nucleus, pontine and medullary reticular formation and vestibular nucleus. The tract arising from the red nucleus is the **rubrospinal tract** but its physiological importance is not yet well understood. The tracts arising from the reticular formation are the **pontine reticulospinal tract** and **medullary reticulospinal tract**. They have a key role in movements which do not require balance or dexterity. The **vestibulospinal tract** originates in the lateral vestibular nucleus and is involved in the control of balance and posture.

1.6 Peripheral nervous system

The peripheral nervous system will now be considered. The afferent division has been discussed already in relation to the ascending nerve tracts originating in the sensory receptors.

Thirty-one pairs of spinal nerves, together with the 12 pairs of cranial nerves referred to earlier, comprise the peripheral nervous system.

1.6.1 THE CRANIAL NERVES

The functions of the cranial nerves are summarized in table 1.3.

Table 1.3 The cranial nerves

Cranial nerve	Sensory	Motor	Autonomic (parasympathetic)
I Olfactory	+		
II Optic	+		
III Oculomotor		+	+
IV Trochlear		+	
V Trigeminal	+		
VI Abducent		+	
VII Facial	+	+	+
VIII Vestibulocochlear	+		
IX Glossopharyngeal	+		+
X Vagus			+
XI Accessory		+	
XII Hypoglossal		+	

1.6.2 THE SPINAL NERVES

The spinal cord is divided into segments from which 31 pairs of spinal nerves arise. There are eight pairs of cervical nerves, 12 thoracic nerves, five lumbar nerves, five sacral nerves and one coccygeal nerve. The peripheral distribution of the spinal nerves is shown in figure 1.10.

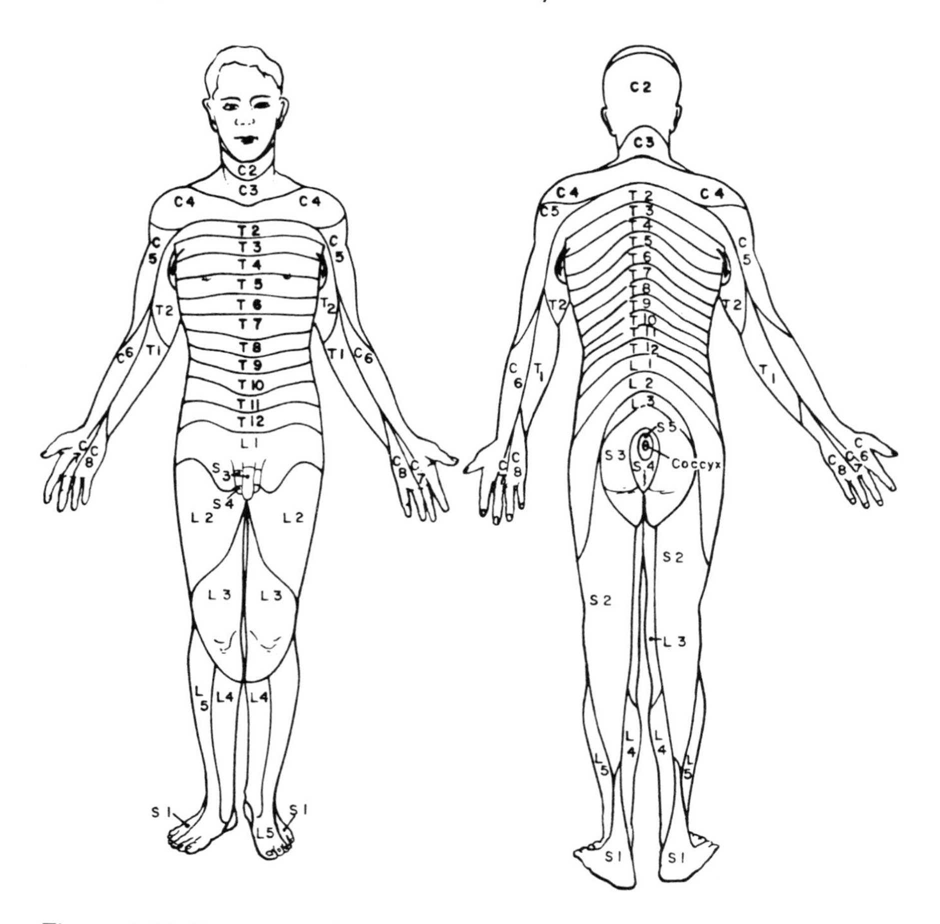

Figure 1.10 Cutaneous distribution of spinal nerves. Reproduced with permission from Barr and Kiernan (1983), p. 79.

Somatic reflex arc (see also section 3.2.2)

Afferent fibres enter the dorsal (posterior) side of the spinal cord via the dorsal root. However, the efferent fibres leave the spinal cord on the ventral (anterior) side via the ventral root. The two roots combine near the spinal cord to give one single spinal nerve.

Inside the grey matter of the spinal cord the sensory nerves may form reflex arcs either directly, or via interneurones, with efferent motor nerves. However, many sensory impulses pass up the spinal cord along neurones located in sensory nerve tracts, as indicated earlier.

The somatic efferent fibres (which supply skeletal muscle) are single motor neurones with cell bodies in the central nervous system, their axons ending without synapse at the motor end-plate.

1.6.3 THE AUTONOMIC NERVES

The autonomic nervous system is an efferent system of nerve fibres which supply the viscera. Unlike skeletal muscle, visceral tissues such as the heart, intestines and uterus display automatic excitability *in vitro* although *in vivo* it may be possible to influence the normally automatic excitability of some visceral tissues. This applies to the heart, the rate of contraction of which may be influenced by cerebral command during activities such as yoga or meditation. An important difference between somatic and autonomic nerve fibres which has been referred to earlier is that firing of the somatic fibres always results in excitation (muscle contraction) whereas firing of an autonomic efferent visceral fibre may result in either excitation or inhibition, depending on the fibre in question. A second important difference between somatic and visceral fibres is that the somatic fibres reach their end organ without synapse whereas there are always at least two fibres involved in the autonomic nervous system, a preganglionic fibre and a postganglionic fibre. A ganglion is a mass of neuronal cell bodies and nerve terminals which acts as a relay station where pre- and postganglionic neurones make connections. A third neurone is found in the **enteric nervous system** in nerves supplying the wall of the alimentary canal.

The autonomic spinal reflex arc

The **autonomic spinal reflex arc**, like the somatic spinal reflex arc, typically involves three neurones. The first is an afferent neurone with its cell body in the dorsal root ganglion. In the grey matter of the spinal cord this neurone connects with a myelinated B nerve fibre, the **preganglionic nerve fibre**, so called because its nerve terminal lies in a ganglion, either close to the spinal cord in the chain of sympathetic ganglia or in a more peripheral ganglion (see pages 28–30). The nerve fibre which leaves the autonomic ganglion is a non-myelinated C fibre, called the **postsynaptic nerve fibre**. Descending nervous activity from the limbic system, hypothalamus and motor centres in the brain influence the autonomic nerve fibres in the same way that nerve traffic descending the spinal cord may influence the final motor outflow along the somatic fibres.

The sympathetic autonomic nervous system

The sympathetic fibres of the autonomic nervous system leave the spinal cord in the 12 thoracic (T) and upper three lumbar (L) segments. The distribution of the sympathetic fibres from each spinal segment displays considerable overlap. However, they can be approximately categorized as supplying the smooth muscle, sweat glands and viscera in the following regions: neurones leaving T1 supply the head, those from T2 supply the neck, those from T3 to T6 supply the thorax, those from T7 to T11 supply the abdomen, and the remainder, from T12 to L3, supply the legs (figure 1.11).

Either side of the spinal cord are a pair of sympathetic chains, the vertical chains of sympathetic ganglia. They are joined to the spinal nerves by white **rami communicantes**. When the sympathetic autonomic fibres leave the spinal cord, they pass through a white ramus communicans to the appropriate ganglion.

Depending on the fibre in question, the preganglionic neurone may:

(a) synapse in the ganglion and re-enter the spinal nerve via a grey ramus communicans;
(b) synapse in the ganglion but then pass directly to the target cell in a sympathetic autonomic nerve;
(c) pass through the sympathetic chain without synapse to the coeliac, superior mesenteric or inferior mesenteric ganglion (figure 1.12).

The parasympathetic autonomic nervous system

The parasympathetic fibres of the autonomic nervous system leave the brainstem in cranial nerves III, VII, IX and X and in nerves leaving sacral segments 2, 3 and 4. Cranial nerves III, VII and IX supply the head, cranial nerve X (the vagus nerve) supplies the thorax and abdomen, and those leaving the sacral segments supply the pelvis (figure 1.11).

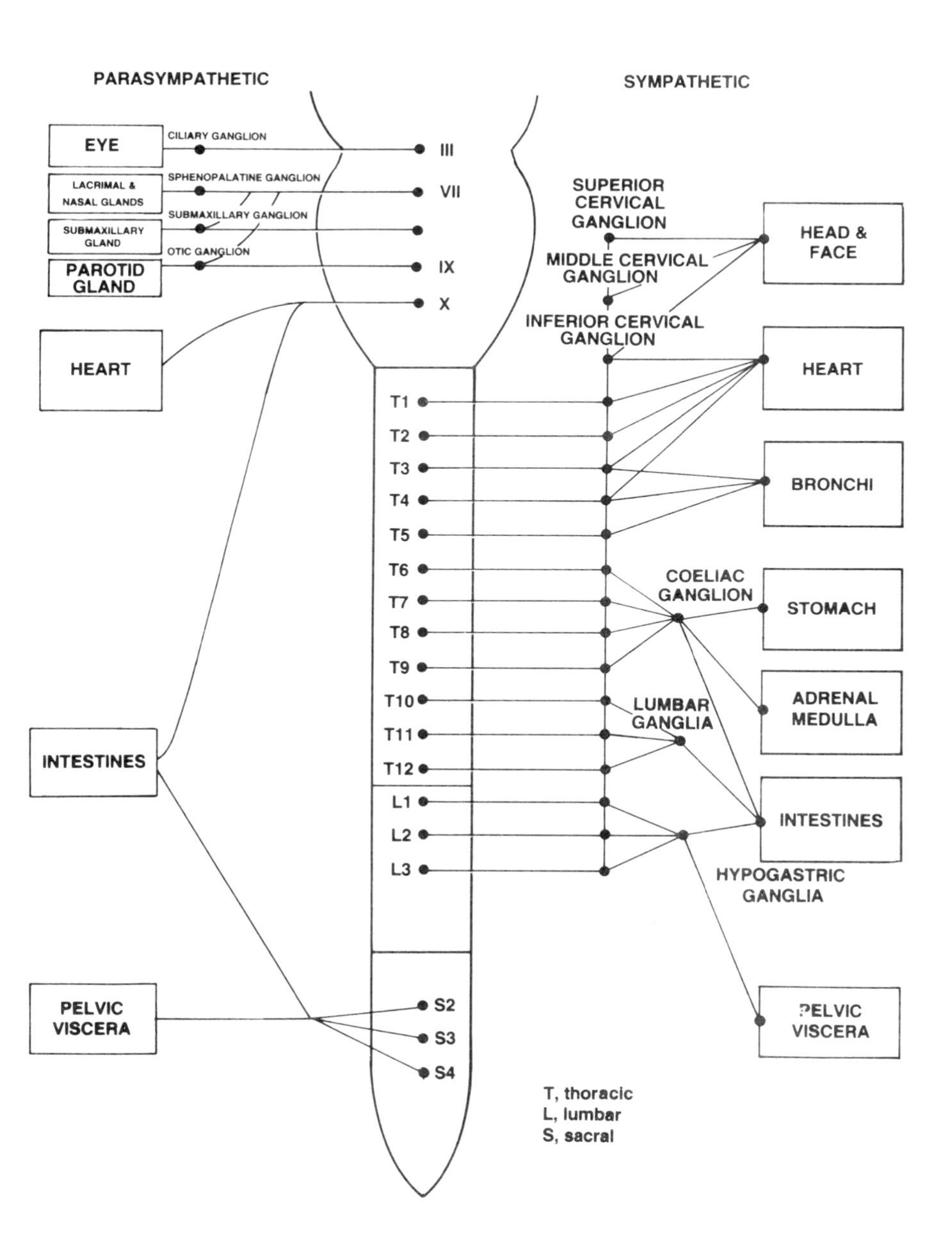

Figure 1.11 Organization of the autonomic nervous system.

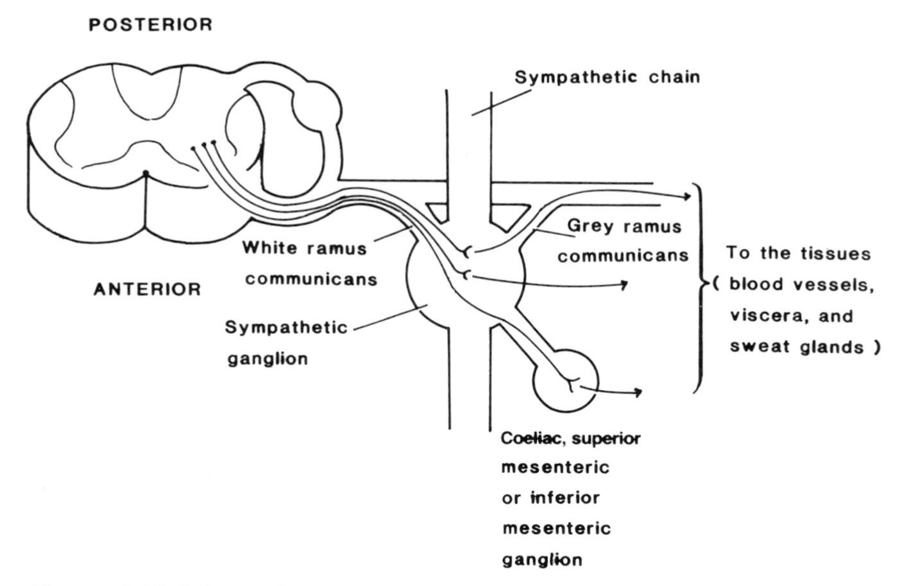

Figure 1.12 Schematic arrangement of the sympathetic division of the autonomic nervous system. Figure adapted from Ganong, W.E. (1977) *Review of Medical Physiology*, 8th edition, Lange, Los Altos, p. 149.

The ganglia of the parasympathetic fibres are close to the end organ; thus the postganglionic axons are short compared with most of those in the sympathetic division of the autonomic nervous system.

Central control of autonomic nervous activity Regions of the medulla, pons and diencephalon are involved in the control of autonomic nervous activity. For example, arterial blood pressure, heart rate and respiration are controlled by neurones in the lower brainstem (medulla and pons), and mechanisms of body temperature regulation are controlled by hypothalamic neurones (in the diencephalon).

The physiology of the autonomic nervous system will be discussed in more detail in the following chapter.

References and further reading

Barr, M.L. and Kiernan, J.A. (1983) *The Human Nervous System: An Anatomical Viewpoint*, 4th edition, Harper and Row, Philadelphia.

Eugene, B.N. and Cherniak, N.S. (1987) Central chemoreceptors *J. Appl. Physiol.* **62**, 389–402.

Lindsley, D.F. and Holmes, J.E. (1984) *Basic Human Neurophysiology*, Elsevier Science, New York.

Ottoson, D. (1983) *Physiology of the Nervous System*, Macmillan, London.

Pick, J. (1970) *The Autonomic Nervous System: Morphological, Comparative, Clinical and Surgical Aspects*, J.B. Lippincott, Philadelphia.

Chapter 2

Physiology of the autonomic nervous system

The autonomic nervous system is not under voluntary control. Its major role is to ensure homeostasis, that is, a stable internal environment. The body's internal environment is monitored by sensory receptors (section 1.5.3) which transmit neural impulses along afferent nerves to the central nervous system. The central nervous system in turn relays the appropriate responses to the incoming sensory information via efferent autonomic nerves. Thus, when the body's internal environment is challenged, the autonomic nervous system responds through its connections with the tissues. In addition, neural activity in the higher centres (such as conscious thought arising in the cerebral cortex or emotions arising in the limbic system) influences brainstem nuclei which form neural connections with sympathetic autonomic neurones. This anticipatory increase in autonomic nervous activity is an important response prior to exercise. In preparing the body for exercise the effects of the sympathetic autonomic nerves help to reduce the challenge that exercise itself presents.

2.1 Neurotransmission

The neurotransmitters of autonomic nerve fibres are summarized in figure 2.1. All preganglionic nerve fibres of the autonomic nervous system (whether sympathetic or parasympathetic) release **acetylcholine** as their neurotransmitter. Postganglionic neurotransmission by acetylcholine occurs in some sympathetic and in all parasympathetic fibres. Most sympathetic effects are mediated by neurones which release noradrenaline as the neurotransmitter at the

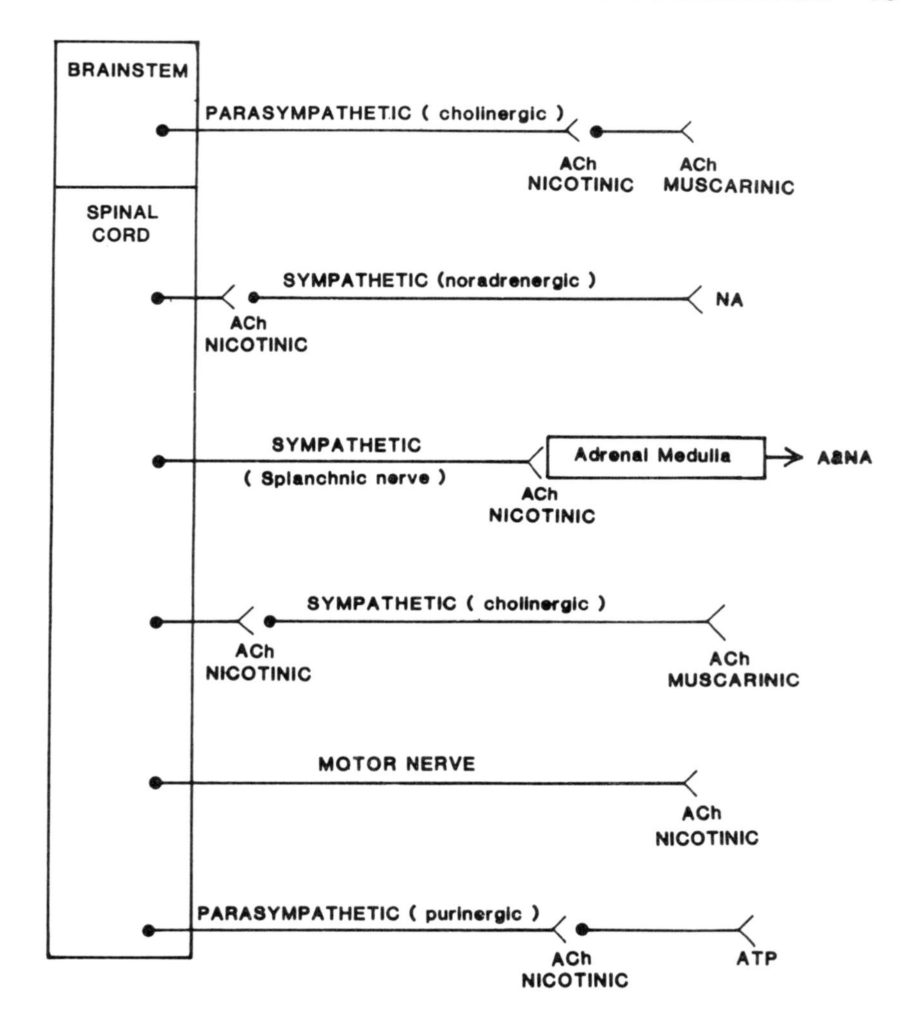

Figure 2.1 Neurotransmission in peripheral nerves.

effector site. These neurones are termed **noradrenergic**. Adrenaline and noradrenaline released from the adrenal medulla are hormones rather than neurotransmitters since they circulate in the blood to their target organs. Various non-cholinergic, non-noradrenergic neurones have also been identified. The most common of these are post-ganglionic neurones which release ATP at the effector site where they mediate relaxation in smooth muscle. These neurones have been termed **purinergic**.

There are five principal stages of chemical transmission:

1. manufacture of the neurotransmitter
2. storage of the neurotransmitter
3. release of the neurotransmitter
4. action of the neurotransmitter
5. deactivation of the neurotransmitter

These five stages are outlined for noradrenaline and for acetylcholine.

2.2 Noradrenaline

2.2.1 MANUFACTURE

Noradrenaline is formed in both peripheral and central neurones as well as in the **chromaffin** cells of the adrenal medulla and other extraneural chromaffin tissue. Noradrenaline (and adrenaline) are derived from the amino acid precursors, tyrosine and phenylalanine which contain the benzene ring characteristic of the aromatic amino acids (figure 2.2).

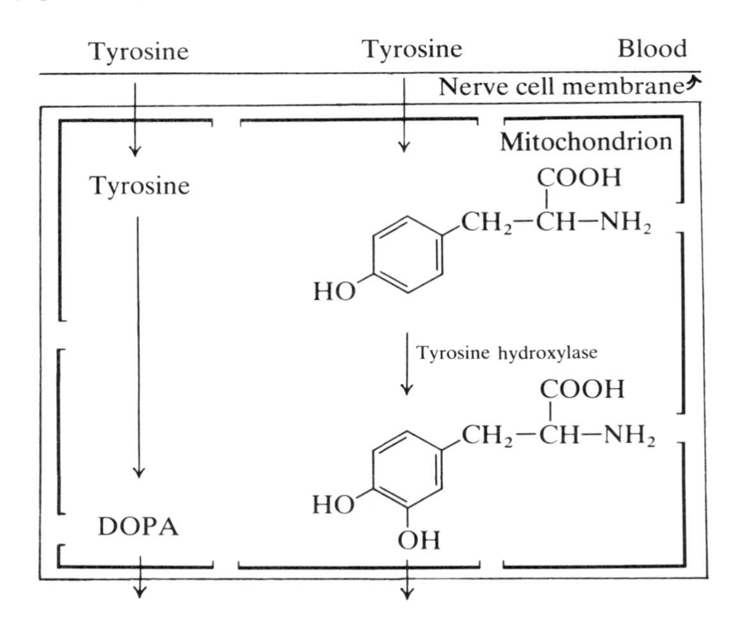

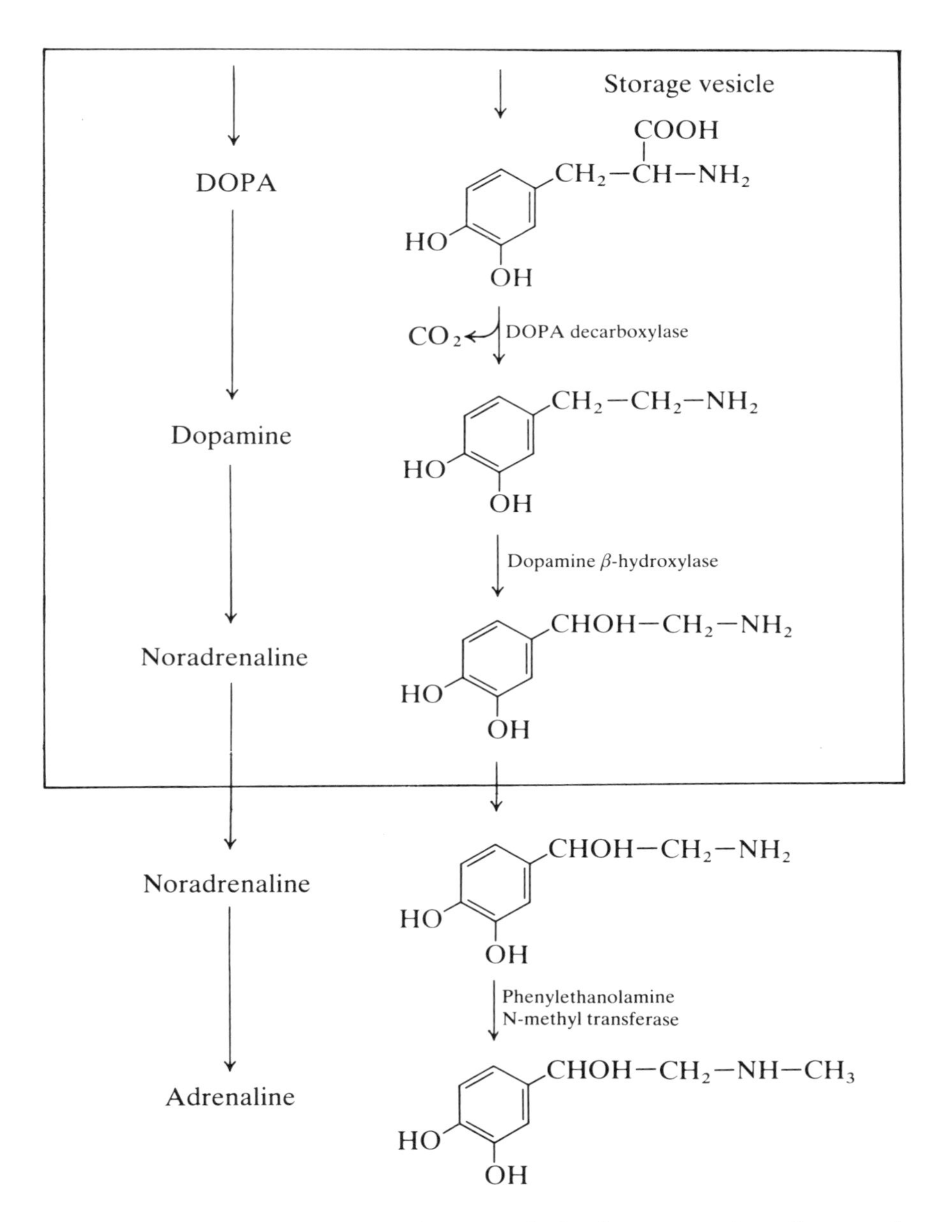

Figure 2.2 Manufacture of the catecholamines noradrenaline and adrenaline.

Manufacture of noradrenaline begins with the uptake of tyrosine into the mitochondria of the nerve cell where it is oxidized to **dihydroxyphenylalanine (dopa)** by **tyrosine hydroxylase**, the rate-limiting reaction of the pathway. Dopa then enters the storage vesicles where it is decarboxylated (carbon dioxide, and thus the acidity, is removed) by dopa decarboxylase to form **dopamine**. Dopamine acts as a neurotransmitter in its own right in some central neurones. However, in peripheral noradrenergic neurones, dopamine is converted to noradrenaline by the enzyme **dopamine β hydroxylase**. In the adrenal medulla noradrenaline leaves the storage vesicles and enters the cytoplasm where it is methylated (a methyl group, CH_3, is added) to adrenaline. The formation of adrenaline outside the storage vesicle is necessary because the enzyme **phenylethanolamine N-methyl transferase**, which catalyses adrenaline synthesis, is located in the cytoplasm. Once formed adrenaline enters the storage vesicle.

2.2.2 STORAGE

The storage vesicles for noradrenaline in the nerve terminal are 40–50 nm in diameter. In the adrenal medulla both catecholamines are stored in vesicles (or granules) that are 0.1–0.5 μm in diameter. In both the neurones and the chromaffin cells of the adrenal medulla in catecholamines are stored in complex form with ATP. A unit of catecholamines and ATP in a molecular ratio of 4:1 forms the contents of one vesicle.

2.2.3 RELEASE

Noradrenaline release from the postganglionic nerve terminals is stimulated by acetylcholine release at the preganglionic nerve terminals. Once released from the noradrenergic nerve terminal, noradrenaline diffuses across the synaptic cleft which is 20–100 nm in diameter. There it reacts with specialized receptors on the postsynaptic membrane (figure 2.3).

In the case of the adrenal medulla the trigger for catecholamine release is stimulation of the splanchnic nerve which, as described in Chapter 1, supplies the adrenal medulla. Since the splanchnic nerve is preganglionic (the adrenal medulla being a modified ganglion), its neurotransmitter is acetylcholine. Catecholamines released from the

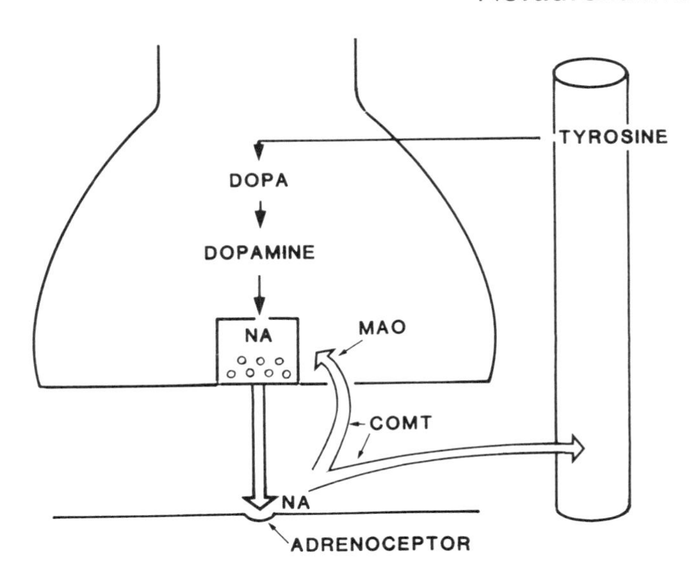

Figure 2.3 Noradrenergic nerve terminal. COMT = Catechol-*o*-methyl transferase; NA = noradrenaline; MAO = monamine oxidase.

adrenal medulla are transported in the blood to receptors in the cell membranes of their target tissues.

2.2.4 ACTION

Catecholamine action is mediated by specialized adrenoceptors in the cell membrane which are classified as α or β. These receptors are further subdivided into two further classes, α_1 and α_2 and β_1 and β_2, as a result of their affinity for pharmacological agents. In addition the α_1- and α_2-adrenoceptors can generally be classified by their location, α_2 being typically located on the presynaptic membrane whereas α_1-adrenoceptors are usually found on the postsynaptic membrane. Adrenaline and noradrenaline are equally potent with regard to the α- and β_1-receptors whereas adrenaline is more potent than noradrenaline with regard to the β_2-receptors (figure 2.4).

Adrenoceptor density is not constant. When receptor stimulation by adrenergic compounds is increased, the number of adrenoceptors is reduced (so-called **down-regulation**). Conversely, when adrenoceptor stimulation is reduced, the adrenoceptor density is

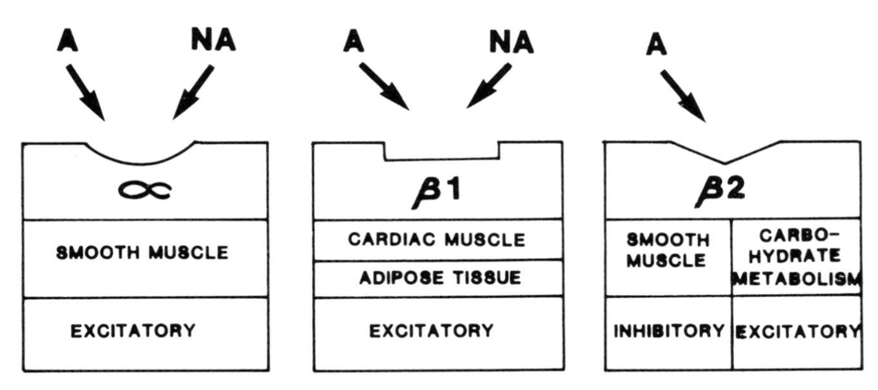

Figure 2.4 Catecholamine affinities for adrenoceptors and predominant effects of stimulation.

increased. The loss of receptors seems to be due to their being internalized in the cell membrane rather than being degraded.

In general, noradrenergic nerves are primarily concerned with cardiovascular responses whereas adrenaline is primarily concerned with β-adrenoceptor-mediated effects on metabolism. The predominant targets of sympathoadrenal activity during exercise are shown in table 2.1.

The classification of receptors in this way (α or β) may not truly represent their nature. There are situations (such as intestinal relaxation) when both α- and β-receptors are involved or where in one condition a given response is mediated by α-receptors and in another condition the same response is mediated by β-receptors. This would be consistent with the view that only one receptor is involved and that this receptor is capable of expressing different (α or β) properties, but not with the view that α- and β-adrenoceptors have different compositions. Further evidence for the latter comes from studies of coronary adrenoceptors, which display temperature-dependent characteristics with α-adrenoceptor effects seen at low temperatures and β-adrenoceptor effects at high temperatures.

Mechanism of catecholamine action

The β-adrenoceptors mediate their effects by activating adenylate cyclase to bring about an increase in **cyclic AMP (cAMP)** which acts as a chemical messenger within the target cell. As noradrenaline and

Table 2.1 Effects of autonomic nervous activity on selected tissues

Organ	Sympathetic		Parasympathetic
	Noradrenergic	Cholinergic	
Heart	↑ heart rate (β)	–	↓ heart rate
	↑ ventricular contractility (β)	–	–
Blood vessels:			
coronary	constriction (α)	–	dilatation
	dilatation (β)	–	–
Skin	constriction (α)	–	–
Skeletal muscle	constriction(α)	–	–
	dilatation (β)	dilatation	–
Abdominal viscera	constriction (α)	–	–
	dilatation (β)	–	–
Lung:			
bronchial muscle	relaxation (β)	–	contraction
Stomach/intestine	↓ motility and tone (β)	–	↑ motility and tone
Eccrine sweat glands	secretion (α)	secretion	–
Liver	↑ glycogenolysis (β/α)	–	↑ glucose storage/ glycogen synthesis
Muscle	↓ insulin-mediated glucose uptake (β)	–	–
Adipose tissue	↑ lipolysis (β)	–	–
	↓ lipolysis (α)	–	–
Pancreas			
β cell	↓ insulin secretion (α)	–	–
	↑ insulin secretion (β)	–	↑ insulin secretion
α cell	↑ glucagon secretion (β)	–	–

adrenaline are the primary chemical messengers, cAMP is known as a second messenger. Cyclic AMP is formed from ATP in the effector cell membrane by the reaction:

$$\text{ATP} \xrightarrow{\text{adenylate cyclase}} \text{cAMP} + \text{ADP}$$

This is followed by a cascade of enzymatic reactions which ultimately results in a specific response. This mechanism occurs in the β-adrenoceptor-mediated breakdown of glycogen (glycogenolysis) and triacylglycerol (lipolysis). The final enzymatic reaction in these cases is modification of the activity of phosphorylase or triacylglycerol lipase, respectively (figure 2.5).

This mechanism plays an important role in the anticipatory stage of exercise. However, in the case of glycogenolysis, the cAMP system is bypassed during exercise itself. In this situation the Ca^{2+} response to motor nerve activity stimulates inactive phosphorylase b kinase leading to increased glycogen breakdown without any contribution from the cAMP system.

In contrast to the increase in intracellular cAMP which occurs in β-adrenoceptor action, the α_2-adrenoceptors inhibit adenylate cyclase, thereby lowering intracellular cAMP to exert their physiological effects.

The α_1-adrenoceptors display quite a different mode of action. When an agonist binds to the α_1 adrenoceptor, **phospholipase C phosphodiesterase** is activated causing an increase in Ca^{2+} ions which bind to the regulatory protein, **calmodulin**. The Ca^{2+}–calmodulin complex then acts as the second chemical messenger to bring about α_1-adrenoceptor-mediated action. For a review of α-adrenergic mechanisms see Exton (1985).

2.2.5 DEACTIVATION

Most of the noradrenaline released from the nerve terminals is taken back up again by the nerve, but some is degraded by one of two enzymes. **Catechol-*o*-methyl transferase (COMT)** is found on the postsynaptic membrane and deactivates both neural and adrenal medullary noradrenaline (and adrenaline). By contrast **monoamine oxidase (MAO)** is found in the mitochondria of noradrenergic neurones and deactivates free noradrenaline in the neurone, but has no effect on noradrenaline in the storage vesicles.

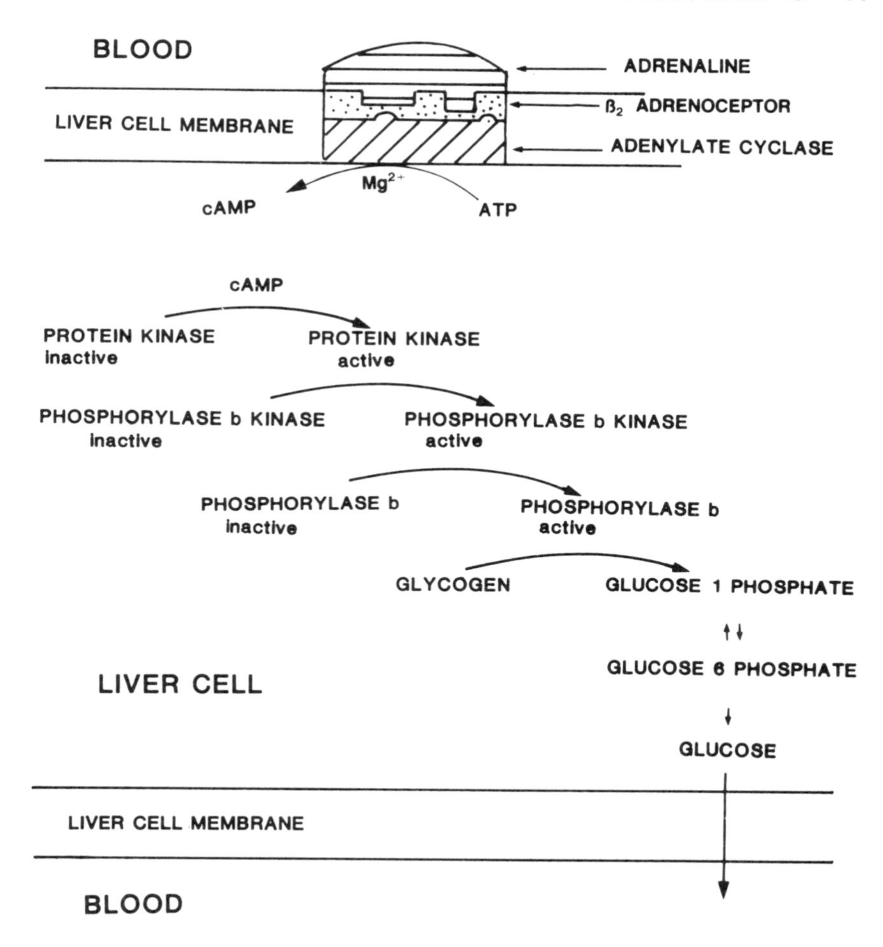

Figure 2.5 Cyclic AMP mediated enzyme cascade during glycogenolysis.

Some of the noradrenaline released from nerve terminals enters the bloodstream (figure 2.3) where in high concentrations (> 10 nmol litre^{-1}) it can exert effects on the cardiovascular system and metabolism and thus may be considered to be acting as a hormone. A small amount of the noradrenaline released (about 2%) is eventually excreted in the urine.

It should be remembered that the main function of noradrenaline is as a neurotransmitter rather than as a hormone. The noradrenaline concentration at the nerve terminal, where it is released, will exceed that found in the blood. Under normal conditions the blood level of noradrenaline is too low to be likely to cause direct physiological effects.

2.3 Adrenal medulla

About four times more adrenaline than noradrenaline is stored in the adrenal medulla. Once released, the catecholamines enter the bloodstream where they act as hormones causing widespread effects. The effects of adrenal medullary catecholamines take a relatively longer period of time to become manifest, and their effects last longer than those mediated by noradrenergic nerves. Adrenaline is about ten times more potent as a hormone than noradrenaline with levels as low as 0.3 nmol $litre^{-1}$ producing physiological effects.

Studies with rats and mice have shown that the relative importance of sympathetic noradrenergic activity and the adrenal medullary secretion of catecholamines depends on the condition to which the animal is exposed. For example, during hypoglycaemia or acute ischaemic trauma, sympathetic nervous activity is suppressed but the adrenal medullary response is enhanced. In the case of trauma there is a biphasic response of the sympathetic nervous system with an increase in sympathetic activity occurring several days after the trauma.

2.4 Assessment of sympathoadrenal activity

There are several ways of deriving information about sympathoadrenal activity and these are by assessing the physiological responses, the effects of pharmacological compounds and of surgery, tissue turnover of noradrenaline, urinary production of catecholamines and their metabolites, plasma catecholamine concentration, and activity in peripheral nerves.

2.4.1 PHYSIOLOGICAL RESPONSES

The oldest way to assess sympathoadrenal activity is to measure the changes in physiological variables that are known to be influenced by autonomic neural activity. For example, significant relationships between plasma catecholamines and haemodynamic responses (heart rate and systolic blood pressure) are seen during isometric handgrip exercise, a cold pressor test and maximal exercise using a bicycle ergometer. However, a major limitation in deriving information about the sympathoadrenal system from haemodynamic measurements during exercise is that there is a concomitant withdrawal of vagal tone which contributes to the overall response. This theme is developed in Chapter 5.

2.4.2 PHARMACOLOGICAL AND SURGICAL METHODS

In order to assess the magnitude of involvement of different arms of the autonomic nervous system it is possible to block the receptors with pharmacological compounds (or, in animal models, to cut specific nerves) and compare the response to a physiological challenge before and after receptor antagonism (or nerve section). Sympathectomized patients (who have had part of their sympathetic nervous system surgically excized) have been extensively studied and compared with individuals with an intact sympathetic nervous system. Any differences between the control and the blocked or denervated state can often be attributed to autonomic activity. However, the effects of denervation may be masked if other systems take over. For example, following sympathectomy the renin/angiotensin system maintains blood pressure in the absence of sympathetic nervous activity, and the ensuing increase in catecholamine sensitivity and increase in adrenal medullary release of catecholamines result in augmented catecholamine action despite the sympathectomy.

2.4.3 TISSUE TURNOVER

In order to look at the sympathetic nervous activity in an individual organ in animals, it is possible to measure tissue turnover of noradrenaline. This involves infusing radiolabelled noradrenaline into a vein to produce a steady state noradrenaline concentration. The experimental intervention is then imposed on the subject. Blood is subsequently sampled and any decrease in plasma radiolabelled noradrenaline is due to increased release of noradrenaline by the tissue where blood is sampled. The calculations also require the measurement of blood flow and arterial samples, yet even with all these data the calculations are controversial. As plasma clearance of catecholamines is generally fairly constant, it is frequently assumed that catecholamine production can be assessed from the circulating level in the blood.

2.4.4 URINARY PRODUCTION OF CATECHOLAMINES

In the case of sympathoadrenal activity it is possible to measure urinary excretion of catecholamines and their metabolites. However, urinary excretion rates only give information about catecholamine activity over blocks of time. Nevertheless, urine measurements of catecholamines can provide a very good overall measure of sympathoadrenal activity.

2.4.5 PLASMA CATECHOLAMINE CONCENTRATION

Sympathoadrenal activity can be assessed from measurements of circulating catecholamines. In resting supine man, blood sampled from a free-flowing vein in the forearm contains 0.05–0.38 nmol litre^{-1} free adrenaline in the plasma and 1.0–5.0 nmol litre^{-1} free noradrenaline in the plasma. It is conventional to measure free catecholamines (whether in venous or arterial blood). However, most of the circulating catecholamines are in conjugated form, mainly as sulphates (that is, sulphur is attached to the hydroxyl group on the catecholamine molecule). In venous blood approximately 70% of adrenaline and 85% of noradrenaline are conjugated in supine resting man. Since the sulphate-conjugated levels of catecholamines do not increase in response to physiological stimulation (postural changes, cold immersion or isometric handgrip exercise), free catecholamine concentration gives the better index of sympathoadrenal activity. However, non-conjugated catecholamines are also bound to plasma proteins and so some secreted material does not appear in a free, unconjugated form.

Although measurements of circulating catecholamines are convenient, the assays and interpretation of the data present problems. The main questions to ask are:

1. How reliable is the assay?
2. Should arterial or venous blood be sampled?
3. Which vessel is the blood taken from?
4. Which posture is the subject in?
5. What time of day is the sample taken?

These points will now be considered.

Reliability of the assay

Catecholamines are difficult to detect in plasma because they are found in such small quantities. Consequently, there is considerable variation between methods and between laboratories in the measurement of catecholamines from the same plasma sample. Fluorimetric assays give highly variable results which are unacceptable. Radioenzymatic or high performance liquid chromatography (HPLC) assays are preferable.

Arterial or venous blood and sampling site

It is preferable to measure catecholamines in arterial blood rather than venous blood since tissue extraction and noradrenaline outflow cannot be assumed to be constant. For example, isometric handgrip exercise leads to a 2.5-fold increase in arterial plasma adrenaline concentration but only a 1.5-fold increase in venous plasma draining the exercising arm compared with the resting arm. In resting individuals the arterial adrenaline levels are about 30–50% higher than the venous levels, indicating there is an extraction of adrenaline of up to 50% by the forearm tissues. In addition, different tissues release different amounts of noradrenaline into the blood. In resting healthy individuals the lungs contribute the greatest proportion of noradrenaline released into venous blood (33%), followed by the kidneys (22%) and skeletal muscle (20%) with other tissues each contributing less than 10% (figure 2.6). Thus noradrenaline measured in blood drawn from an accessible vein (such as an antecubital vein) may not provide a reliable index of overall sympathetic nervous activity, and therefore venous blood should not be sampled. Sampling blood from an artery is the ideal site, but this is a specialized clinical procedure and an alternative solution is to sample from a dorsal hand vein, with the hand heated to increase blood flow through the arteriovenous anastamoses. In this way arterialized venous blood is sampled.

Posture

Moving from a lying or sitting position to standing leads to a number of cardiovascular reflex responses which serve to maintain arterial blood pressure. These include an increase in systemic vascular resistance which is mediated by noradrenergic nerves. Evidence for

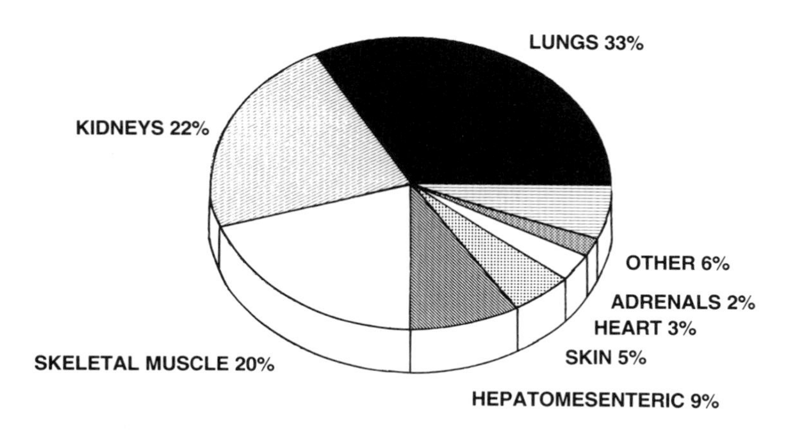

Figure 2.6 Tissue contribution to total noradrenaline release in resting man. Data from Esler *et al.* (1984).

this comes from direct measurements of sympathetic activity in the muscle of subjects lying, sitting or standing.

Time of day

Postural changes, together with sleep, give rise to a circadian rhythm of plasma noradrenaline. Plasma noradrenaline has a trough at around 02.20 h and plasma adrenaline has a trough at around 03.20 h. However, the diurnal rhythm of adrenaline has no bearing on sleep or posture, and is therefore probably controlled by a circadian oscillator, the likely location of which is the suprachiasmatic nucleus.

In addition to the above circadian rhythms, plasma catecholamines display ultradian (rapid) rhythms. This could explain why the reproducibility of resting or exercise values in a single individual is not possible even when the variability of the catecholamine assay is accounted for. Measurements repeated on two days within the same 2-hour period one week apart vary by up to 50% for noradrenaline and by up to 37% for adrenaline. One should be cautious about relying on group data (although they are reproducible) in view of the fact that the mean values mask large intra-individual variability. Any claim for the effect of interventions, such as exercise or training, should be demonstrable in individual subjects.

Blood threshold concentrations for catecholamine action Several physiological and pathophysiological states induce such a high spillover of noradrenaline at the nerve terminal that it is in a sufficiently high concentration in the blood to act as a hormone. With the exception of phaeochromocytoma (catecholamine-producing tumour), a myocardial infarction produces the highest level of circulating noradrenaline, with values as high as 50–60 nmol $litre^{-1}$ immediately post-infarction. In healthy individuals the highest levels of circulating adrenaline are seen during vigorous exercise with levels of up to 20 nmol $litre^{-1}$ being seen following the Wingate test of anaerobic power. A high concentration of plasma noradrenaline is also seen in patients with ketoacidosis and during recovery from surgery. Plasma adrenaline levels are highest in phaeochromocytoma and myocardial infarction, followed by ketoacidosis and then vigorous exercise.

Studies using infusions of adrenaline and noradrenaline have shown that there are tissue-specific threshold levels of plasma catecholamines that are required to produce a response (figures 2.7 and 2.8). Only a small increase in circulating adrenaline is required to produce an increase in heart rate. In contrast plasma noradrenaline must increase to above 10 nmol $litre^{-1}$ to produce a decrease in insulin secretion from the pancreatic β cells. Plasma adrenaline does not increase during exercise until the work load is as high as 75% of the maximum oxygen consumption. It is therefore not responsible for any of the cardiovascular and metabolic responses to light exercise, and is only of any real importance in severe exercise. Similarly, as a hormone, noradrenaline has no physiological effects until an individual is working hard (above 75% of maximum rate of oxygen consumption) when the plasma level is greater than about 10.0 nmol $litre^{-1}$. Any effects of noradrenaline at lower work intensities are more likely to be due to the action of noradrenaline at the nerve terminal where, as stated earlier, the concentration will be higher than in the blood.

2.4.6 NERVE RECORDINGS

There is a linear relationship between circulating catecholamines and directly recorded neural activity in superficial sympathetic nerves of muscle both at rest and during handgrip exercise in man. The ultradian rhythms of plasma noradrenaline are consistent with the

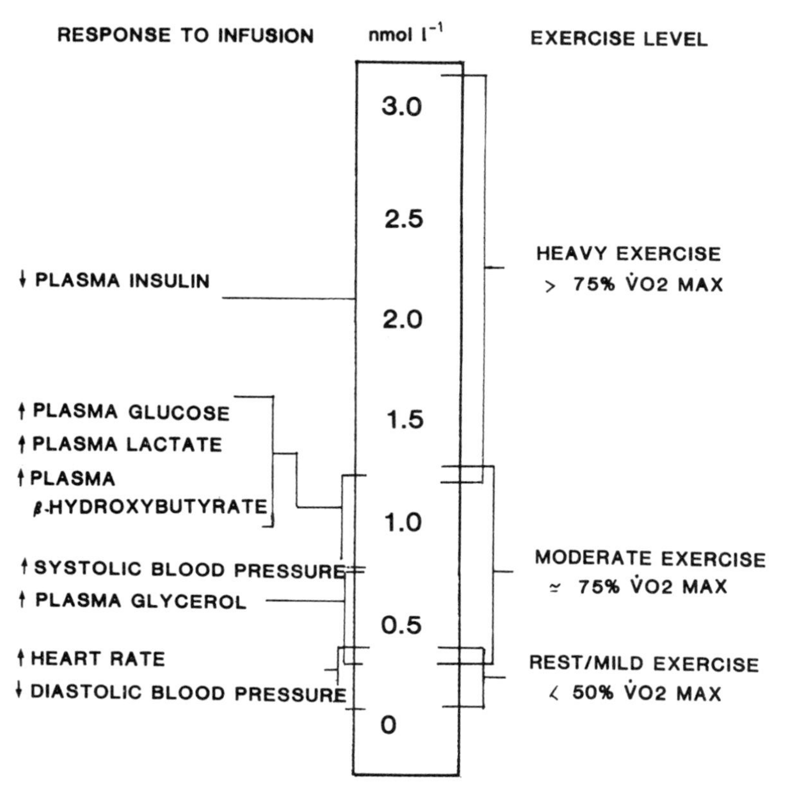

Figure 2.7 Plasma adrenaline thresholds and circulating levels during exercise.

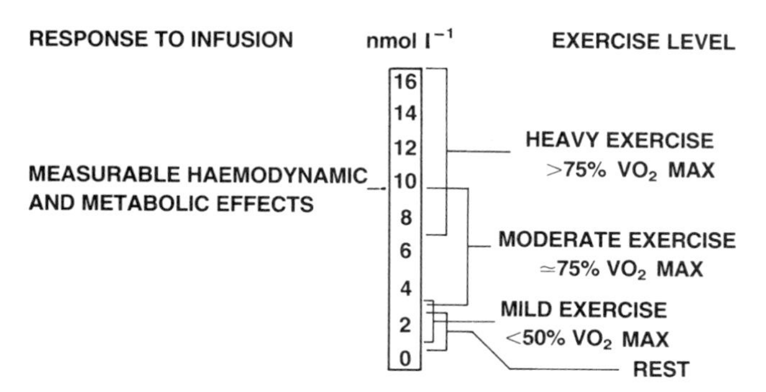

Figure 2.8 Plasma noradrenaline thresholds and circulating levels during exercise.

rapid rhythm of bursts of neural activity recorded directly from median or peroneal muscle nerves. Interestingly, the pattern of these rapid bursts of electrical activity and periods of electrical silence is repeatable in the same individual even when a period of two months has elapsed between recordings (figure 2.9). These direct recordings give absolute measures of sympathetic nervous activity but the method is impractical for many exercising situations.

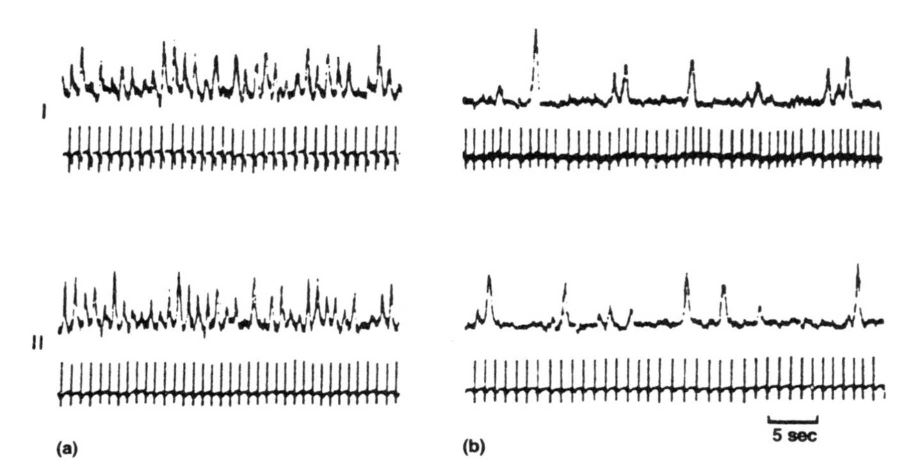

Figure 2.9 Repeated recordings of muscle nerve sympathetic activity. Reproduced with permission from Sundlöf and Wallin (1977). (a) and (b) are two different subjects, with a time interval of 2 months (subject a) and 3 weeks (subject b) between recordings I and II.

2.5 Acetylcholine

2.5.1 MANUFACTURE

Acetylcholine is formed in the mitochondria of the cholinergic nerve fibres. Free choline is taken up from the blood where it reacts with acetyl CoA already present in the mitochondria (figure 2.10). This acetyl CoA is derived from glucose metabolism, which explains why cholinergic nerve function relies so critically on an adequate glucose supply. The enzyme which catalyses this reaction is **choline acetyltransferase** (or simply **choline acetylase**). The rate of

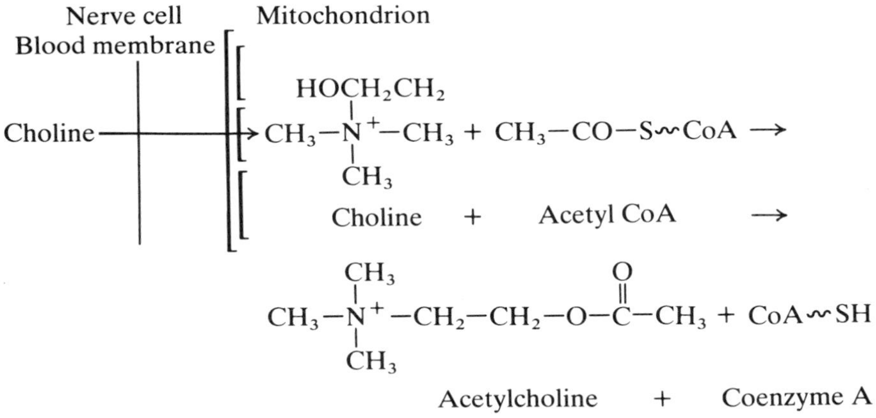

Nerve cell Blood membrane — Mitochondrion — Choline →

$$\mathrm{CH_3{-}\overset{\overset{\displaystyle HOCH_2CH_2}{|}}{\underset{\underset{\displaystyle CH_3}{|}}{N^+}}{-}CH_3 + CH_3{-}CO{-}S{\sim}CoA \rightarrow}$$

Choline + Acetyl CoA →

$$\mathrm{CH_3{-}\overset{\overset{\displaystyle CH_3}{|}}{\underset{\underset{\displaystyle CH_3}{|}}{N^+}}{-}CH_2{-}CH_2{-}O{-}\overset{\overset{\displaystyle O}{||}}{C}{-}CH_3 + CoA{\sim}SH}$$

Acetylcholine + Coenzyme A

Figure 2.10 Manufacture of acetylcholine.

manufacture of acetylcholine is determined by the electrical activity in the cell. For example, in the superior cervical ganglion of the cat, acetylcholine synthesis increases seven-fold during maximum stimulation of the preganglionic nerve fibres.

2.5.2 STORAGE

The manufactured acetylcholine is stored in vesicles about 30–60 nm in diameter. Some of these vesicles are located in the axon, but the main location is the nerve ending. The several hundred molecules of acetylcholine stored in each vesicle form a 'packet' or quantum of acetylcholine.

2.5.3 RELEASE

Some acetylcholine is released at random intervals even in the resting state. However, it is only when the nerves are stimulated that sufficient acetylcholine is released to transmit the impulse to the postsynaptic membrane. When a nerve action potential arrives at the nerve terminal, calcium channels open so that Ca^{2+} ions enter the nerve terminal to bring about release of acetylcholine from the storage vesicles. For this to occur the vesicles fuse with the nerve cell membrane, empty their contents and then collapse. Once released, acetylcholine diffuses across the synaptic cleft, which electron microscope studies have shown to be about 15–50 nm wide.

2.5.4 ACTION

When acetylcholine reaches the postsynaptic membrane, it reacts with specific glycoprotein receptor molecules (figure 2.11). There are three kinds of receptor for acetylcholine. These receptors are the nicotinic and muscarinic receptors of the autonomic nervous system and the receptors at the motor end plate which have some nicotinic properties. In the autonomic nervous system the ganglion cell receptors are nicotinic. The nicotinic receptors can be stimulated by nicotine extracted from the tobacco plant and blocked by hexamethonium.

The smooth muscle, cardiac muscle and gland cell cholinergic receptors are **muscarinic** and can be stimulated by **muscarine**, a fungal extract, and blocked by **atropine**. Motor end-plate receptors have nicotinic properties but cannot be blocked by the pharmacological agents which block the nicotinic receptors of the ganglion cells. They can be blocked by curariform drugs, including curare itself.

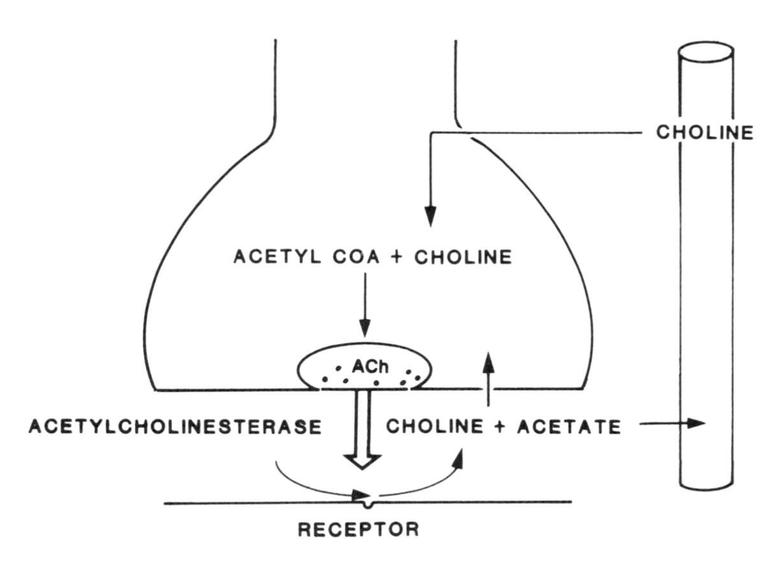

Figure 2.11 Cholinergic nerve terminal.

Acetylcholine has an affinity with the receptor by virtue of its physical and electrical characteristics. When sufficient acetylcholine combines with the receptor, the postsynaptic cell membrane becomes depolarized.

2.5.5 DEACTIVATION

Acetylcholine is very rapidly removed by the action of acetylcholinesterase situated on the cell membrane near the receptor. This enzyme hydrolyses acetylcholine to acetate and choline (figure 2.12). The choline is taken back up into the releasing nerve terminal whereas the acetate diffuses into the bloodstream. The choline that is taken back up is used to form new acetylcholine. In this way the acetylcholine content of the nerve is kept virtually constant, even during neural activity.

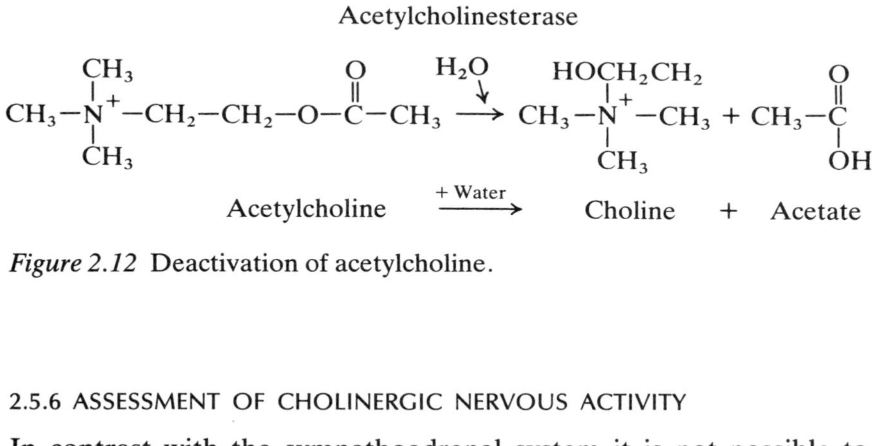

Acetylcholinesterase

$$CH_3-\overset{\overset{CH_3}{|}}{\underset{\underset{CH_3}{|}}{N^+}}-CH_2-CH_2-O-\overset{\overset{O}{\|}}{C}-CH_3 \xrightarrow{H_2O} CH_3-\overset{\overset{HOCH_2CH_2}{|}}{\underset{\underset{CH_3}{|}}{N^+}}-CH_3 + CH_3-\overset{\overset{O}{\|}}{\underset{\underset{OH}{|}}{C}}$$

Acetylcholine $\xrightarrow{+\text{ Water}}$ Choline + Acetate

Figure 2.12 Deactivation of acetylcholine.

2.5.6 ASSESSMENT OF CHOLINERGIC NERVOUS ACTIVITY

In contrast with the sympathoadrenal system it is not possible to measure plasma acetylcholine levels as an index of cholinergic nervous activity. The reason for this is that acetylcholine is deactivated extremely rapidly following its release at the nerve terminal. It therefore does not leak into the bloodstream. Furthermore, there is no cholinergic ganglion comparable to the adrenal medulla which releases a derivative of acetylcholine to act as a hormone. The effects of cholinergic nerves may be deduced using pharmacological compounds which either mimic or inhibit the activity of these nerves, or by assessing the changes in physiological variables which are influenced by cholinergic nerves.

2.6 Non-noradrenergic non-cholinergic neural transmitters

The classical division of the autonomic nervous system into a noradrenergic and a cholinergic component was questioned in the 1960s by the discovery of nerves in the gut and bladder which could not be blocked by either noradrenergic or cholinergic antagonists. It is now recognized that non-noradrenergic, non-cholinergic nerves supply the urogenital tract and the cardiorespiratory system as well as the gastrointestinal tract and bladder in all vertebrates. For an historical review of the development of understanding of non-noradrenergic non-cholinergic neurotransmission see Burnstock (1986a,b). The principal neurotransmitter of these nerves is ATP, and the nerves are termed purinergic. However, many other transmitters have been proposed following more recent immunohistochemical studies. In addition, the non-noradrenergic non-cholinergic compounds are known to act as cotransmitters with noradrenaline or acetylcholine.

References and further reading

Ahlquist, R.P. (1948) A study of the adrenotropic receptors. *Am. J. Physiol.*, **153**, 586–600.

Banks, P., Bartley, W. and Birt, L.M. (1976) *The Biochemistry of the Tissues*, John Wiley, London.

Berthelsen, S. and Pettinger W.A. (1977) A fundamental basis for classification of α-adrenergic receptors. *Life Sci.*, **21**, 58–601.

Burke, D., Sundlöf, G. and Wallin, B.G. (1977) Postural effects on muscle nerve sympathetic activity in man. *J. Physiol. (Lond.)*, **272**, 399–414.

Burnstock, G., Campbell, G., Bennett, M. and Holman, M.E. (1963) Inhibition of the smooth muscle of the taenia coli. *Nature*, **200**, 581–582.

Burnstock, G. (1971) Neural nomenclature. *Nature*, **299**, 282–283.

Burnstock, G. (1986a) The changing face of autonomic neurotransmission. *Acta Physiol. Scand.*, **126**, 67–91.

Burnstock, G. (1986b) The non-adrenergic non-cholinergic nervous system. *Arch. Int. Pharmacodyn.* Suppl. **280**, 1–15.

Christensen, N.J. (1988) Methods of studying sympathoadrenal activity in man. *Acta Med. Scand.*, **223**, 481–483.

Chuang, D.M., Kinnier, W.J., Farber, L. and Costa, E. (1980) A biochemical study of receptor internalisation during β-adrenergic receptor desensitisation in frog erythrocytes. *Mol. Pharmacol.*, **18**, 348–55.

Clutter, W.E., Bier, D.M., Shah, S.D. and Cryer, P.E. (1980) Epinephrine plasma metabolic clearance rates and physiologic thresholds for metabolic and hemodynamic actions in man. *J. Clin. Invest.*, **66**, 94–101.

Cryer, P.E. (1980) Physiology and pathophysiology of the human sympathoadrenal neuroendocrine system. *N. Eng. J. Med.*, **303**, 436–444.

Esler, M., Jennings, G., Leonard, P., Sacharias, N., Burke, F., Johos, J. and Blombery, P. (1984) Contributions of individual organs to total noradrenaline release in humans. *Acta Physiol. Scand.* Suppl., **527**, 11–16.

Exton, J.H. (1985) Mechanisms involved in α adrenergic phenomena. *Am. J. Physiol.*, **248**, E633–E647.

Galbo, H., Holst, J.J. and Christensen, N.J. (1975) Glucagon and plasma catecholamine responses to graded and prolonged exercise in man. *J. Appl. Physiol.*, **38**, 70–76.

Halter, J.B., Pflug, A.E. and Tolas, A.G. (1980) Arterial–venous differences of plasma catecholamines in man. *Metabolism*, **29**, 9–12.

Halter, J.B., Stratton, J.R. and Pfeifer, M.A. (1984) Plasma catecholamines and hemodynamic responses to stress states in man. *Acta Physiol. Scand.* Suppl., **527**, 31–38.

Hedberg, A. (1983) Adrenergic receptors. Methods of determination and mechanisms of regulation. *Acta Med. Scand.* Suppl., **672**, 7–15.

Hjemdahl, P. (1984) Inter-laboratory comparison of plasma catecholamine determinations using several different assays. *Acta Physiol. Scand.*, Suppl., **527**, 43–54.

Hjemdahl, P., Eklund, B. and Kaijser, L. (1982) Catecholamine handling by the human forearm at rest and during isometric exercise and lower body negative pressure. *Br. J. Pharmacol.*, **77**, 324P.

Jörgensen, L.S., Bönlöcke, L. and Christensen, N.J. (1985) Plasma adrenaline and noradrenaline during mental stress and isometric exercise in man. The role of arterial sampling. *Scand. J. Clin. Lab. Invest.*, **45**, 447–452.

Joyce, D.A., Beilin, L.J., Vandorgen, R. and Davidson, L. (1982) Plasma free and sulfate conjugated catecholamine levels during acute physiological stimulation in man. *Life Sci.*, **30**, 447–454.

Levin, B.E. and Natelson, B.J. (1980) The relation of plasma norepinephrine and epinephrine levels over time in humans. *J. Aut. Nerv. System.*, **2**, 315–325.

Levin, B.E., Rappaport, M. and Natelson, B.H. (1979) Ultradian variations of plasma noradrenaline in humans. *Life Sci.*, **25**, 621–628.

Martinson, J. and Muren, A. (1963) Excitatory and inhibitory effects of vagus stimulation on gastric motility in the cat. *Acta Physiol. Scand.*, **57**, 309–316.

Moreland, R.S. and Bohr, D.F. (1984) Adrenergic control of coronary arteries. *Fed. Proc.*, **43**, 2857–2861.

Nyberg, G. (1981) Vagal and sympathetic contributions to the heart rate at rest and during isometric and dynamic exercise in young healthy men. *J. Cardiovasc. Pharmacol.*, **3**, 1243–1250.

Pérronnet, F., Blier, P., Brisson, G., Diamond, P., Ledoux, M. and Volle, M. (1986) Reproducibility of plasma catecholamine concentration at rest and during exercise in man. *Eur. J. Appl. Physiol.*, **54**, 555–558.

Silverberg, A.B., Shah, S.D., Haymond, M.W. and Cryer, P.E. (1978) Norepinephrine: hormone and neurotransmitter in man. *Am. J. Physiol.*, **234**, E252–E256.

Sundlöf, G. and Wallin, B.G. (1977) The variability of muscle nerve sympathetic activity in resting recumbent man. *J. Physiol. (Lond.)*, **272**, E35–E40.

Wallin, B.G., Sundlöf, G., Eriksson, B-M., Dominiak, P., Grobecker, H. and Lindblad, L.E. (1981) Plasma noradrenaline correlates to sympathetic muscle nerve activity in normotensive man. *Acta Physiol. Scand.*, **111**, 69–73.

Wallin, B.G., Mörlin, C. and Hjemdahl, P. (1982) Muscle sympathetic activity and venous plasma noradrenaline concentrations during static exercise in normotensive and hypertensive subjects. *Acta Physiol. Scand.*, **129**, 489–497.

Young, J.B. and Landsberg, L. (1979) Catecholamines and the sympathoadrenal system: the regulation of metabolism, in *Contemporary Endocrinology*, Vol. 1, (ed. S. Ingbar), Plenum Medical Book Company, New York, pp. 245–303.

Young, J.B., Rosa, R.M. and Landsberg, L. (1984) Dissociation of sympathetic nervous system and adrenal medullary responses. *Am. J. Physiol.*, **247**, E35–E40.

Chapter 3

Neuromuscular function

The nervous system plays a complex role in muscular activity, with intricate involvement of central and peripheral neurones. An overview of the relationship between the nervous system and skeletal muscle activity is given in this chapter. Attention is given to skeletal muscle structure and function and the role of sensory and motor nerves in order to contrast the voluntary control of movement by motor control systems with the involuntary control of homeostasis by autonomic nerves.

3.1 Skeletal muscle

3.1.1 STRUCTURE AND FUNCTION

Skeletal muscle is made up of many **fasciculi** (bundles) of muscle fibres (cells). Each fasciculus is enclosed by the **endomysium** which is a form of connective tissue. Several fasciculi are connected by **perimysium**, a further connective tissue. Yet another connective tissue, the **epimysium**, binds all the fasciculi together to form the whole muscle.

The individual fibres are up to 100 μm in diameter and are the length of the whole muscle, which may be as long as a third of a metre. Each muscle cell is innervated by a somatic nerve fibre which has its ending about halfway along the length of the muscle cell. Each muscle fibre comprises a cell membrane, called the **sarcolemma**, which encloses the **sarcoplasm**. The sarcolemma joins with the tendon cells attached to bone at the end of the muscle. About 80% of the interior of the cell comprises myofibrils which are only 1–2 μm in diameter and extend for the full length of the cell. The remaining subcellular constituents include the nuclei, mitochondria, myoglobin, glycogen, lipid, ATP and creatine phosphate.

When a longitudinal section of a muscle fibre is viewed using an electron microscope, a regular pattern of light and dark transverse bands is apparent. These bands are due to the presence of further filaments in the myofibrils which are arranged in segments called **sarcomeres**. The sarcomere is the functional unit of muscle contraction (figure 3.1).

Sarcomeres contain thick and thin protein filaments arranged in a hexagonal manner, with six thin filaments surrounding one thick filament. Under an electron microscope the thick filaments are seen as dark bands, called **A bands**. The light band between the A bands is the **I band**. The thin filaments surrounding the thick filaments are

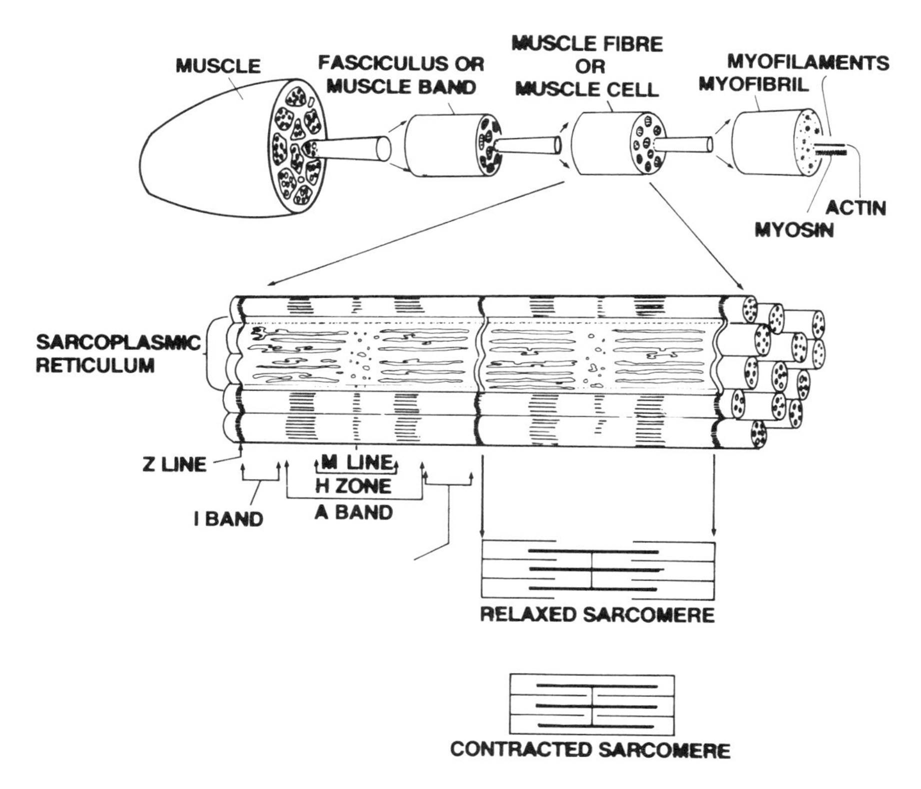

Figure 3.1 Structure of skeletal muscle.

connected to the next sarcomere by the so-called **Z line**. The thin filaments are not continuous along the length of the thick filament; the gap between the points where they end is comprised only of thick filaments and is the **H zone**. In the same way that the thin filaments are held together by the Z line, the thick filaments are held together by the **M line**.

The thick filaments are made of the protein **myosin**. The myosin molecules are arranged in a single chain with their globular heads protruding to form crossbridges. These crossbridges extend out from the tail part of the myosin molecules in every direction in a spiral arrangement. The thin filaments are comprised mainly of the protein **actin** as well as two further proteins, **tropomyosin** and **troponin**, which are involved in regulating muscle contraction. The actin molecules form a double helical chain around which are two double helical chains of tropomyosin molecules. A molecule of **troponin** is bound to both actin and tropomyosin at regular intervals along the thin filament.

During muscular contraction both the H bands and I zones shorten whereas there is no change in the length of the A band (figure 3.1). Thus, the contractile process is associated with the movement of the ends of the actin filaments towards each other until they may finally overlap at the centre of the sarcomere. However, there is an optimal sarcomere length at which muscle tension is maximal. This is the point where the maximum number of myosin crossbridges interact with the actin molecules, and occurs just before the thin filaments overlap in the middle of the sarcomere. The contractile process occurs as a result of an attraction of the myosin crossbridges for specific binding sites on the actin molecules. In the resting state these binding sites are covered by the tropomyosin molecules, held in place by troponin. The binding sites are revealed when ionized calcium (Ca^{2+}) binds to the troponin molecule. The Ca^{2+} ions are released in response to an action potential at the motor end-plate which effectively initiates the contractile process. This account of muscle contraction is called the **sliding filament theory**, a more comprehensive view of which is provided by Pollack (1983).

3.1.2 TYPES OF MUSCLE FIBRE (TABLE 3.1)

Muscle fibres differ with respect to the rate at which they contract in response to nervous stimulation. The slowest of these twitch fibres

Table 3.1 Muscle fibre characteristics

	Fibre type		
	I (red)	IIa (intermediate)	IIb (white)
Time to peak tension	slow	fast	fastest
Mitochondrial density	high	high	low
Capillary density	high	intermediate	low
Myoglobin content	high	high	low
Oxidative enzyme supply	good	good	poor
Glycolytic enzyme supply	poor	intermediate	good
Glycogen content	low	high	highest
Muscle fibre diameter	small	intermediate	large
Rate of fatigue	slow	intermediate	fast

contract in around 110 ms and the fastest in around 50 ms. The proportion of slow (type I) and fast (type II) twitch fibres in a given muscle varies between different muscles in the same individual, and also between individuals for the same muscle. Typically there is a higher proportion of slow twitch fibres in the postural muscles such as the soleus than in the non-postural muscles such as the triceps brachii.

The metabolic capacities of the two fibre types are consistent with the nature of the contraction with which they are associated. The slow twitch fibres are rich in mitrochondria and oxidative enzymes and are well-supplied with capillaries. This facilitates oxidative metabolism, essential for long-term exercise. Thus when the tension produced by muscle contraction is low, these slow twitch fibres are recruited first both in voluntary and reflex activity.

On the basis of metabolic capacity (and not time for contraction) there are two types of fast twitch fibre. The first type (type IIa), like slow twitch fibres, has a good supply of mitochondria and oxidative enzymes and is therefore suited to repeated muscular contractions. In contrast the second type (type IIb) has a high concentration of anaerobic glycolytic enzymes and is poorly supplied with mitochondria and oxidative enzymes. Moreover, the type IIb fibres are larger in diameter than either the type IIa or type I fibres, hence permitting a greater development of tension for a given fibre length.

Thus the type IIb fast twitch fibre is ideally suited for activities which require speed and strength.

As stated above, the proportion of each fibre type in a given muscle varies between individuals. These individual differences are most marked between élite athletes: distance runners tend to have a greater proportion of slow twitch fibres in their leg muscles than either non-athletes or élite sprinters. In contrast the élite sprinter has a greater proportion of fast twitch fibres than either non-athletes or élite distance runners. It is not clear whether fibre types can be altered by training or whether the distribution is entirely genetically endowed.

3.1.3 ENERGY FOR MUSCULAR CONTRACTION

Energy for muscular contraction is derived from adenosine triphosphate (ATP). This is broken down on the myosin crossbridge which contains a specific ATPase. Thus, energy released from the breakdown of ATP to ADP + Pi is temporarily transferred to the myosin crossbridge which is in contact with the actin filament. Energy from the breakdown of ATP pulls the actin filaments towards the centre of the sarcomere. Persistent binding of the myosin crossbridge to the active site of the actin molecule is prevented by the addition of a further molecule of ATP to the myosin crossbridge. As a result the crossbridge is able to bind to a new actin site and again produce movement of the latter molecules towards the centre of the sarcomere. The number of myosin crossbridges which are bound to actin molecules determines the strength of contraction. During passive limb movements the actin filaments move past the myosin filaments freely without the requirement of energy from ATP breakdown.

The myofibrils are covered by the **sarcoplasmic reticulum** which comprises lateral sacs connected by longitudinally running tubules. Additional transverse tubules run perpendicularly to the myofibrils at the border of the A and I bands at the site of the lateral sacs (figure 3.1). The transverse tubules are an important means for conducting the action potential over the entire muscle fibre. As the action potential passes over the surrounding lateral sacs it causes Ca^{2+} release from the lateral sacs with some Ca^{2+} ions also entering from the extracellular compartment. The Ca^{2+} binds to troponin to initiate the contractile process.

When the action potential ceases, Ca^{2+} ions are actively pumped back into the sarcoplasmic reticulum. This active transport of Ca^{2+} requires ATP. The active removal of Ca^{2+} lowers the concentration of the free ion which subsequently causes Ca^{2+} to become unbound from the binding sites on the troponin.

Therefore ATP is essential both for driving muscle contraction and for bringing about relaxation. Since ATP is not stored, except in very small amounts, it must be supplied to the muscle. The rate at which ATP is supplied is a major determinant of the rate at which the muscle cells can produce work. The cells are highly sensitive to change in ATP concentration. ATP is supplied from ADP and phosphate ions, and initially the phosphate ions are derived from creatine phosphate. This supply is shortlived, and after only a few seconds further ATP is derived from the breakdown of muscle glycogen and blood-borne glucose and free fatty acids in the muscle. These processes are controlled to provide ATP in proportion to muscle requirements. This control is provided both by mechanisms intrinsic to the local metabolic pathways and by the sympathoadrenal system, either directly or through its influence on endocrine glands including the pancreas. The role of the autonomic nervous system in the provision of fuel for muscle contraction will be discussed in Chapter 4.

3.1.4 EFFICIENCY OF MUSCULAR WORK

At rest a 70 kg individual expends approximately 5 kJ of energy per minute. During maximal sustained muscular effort this may increase to around 70 kJ per minute. However, the maximum amount of sustained mechanical work that can be done is generally less than 25% of the metabolic energy expenditure, with the remaining energy being lost as heat (Chapter 6). Mechanical efficiency (ME) is calculated using the following equation:

$$\text{ME} = \frac{\text{Mechanical work}}{\text{Total energy expenditure}} \times 100$$

Thus, if an individual carries out 8.8 kJ min^{-1} (900 kg min^{-1}) of work using, for example, a bicycle ergometer and has an energy expenditure of 50 kJ min^{-1} (oxygen consumption is 2.5 litres min^{-1}), his mechanical efficiency would be 17.6%.

$$\mathrm{ME} = \frac{8.8}{50} \times 100 = 17.6\%$$

Mechanical efficiency is determined by the speed of muscular contraction, force of contraction, degree of training and muscular fatigue.

3.2 Control of movement: peripheral mechanisms

3.2.1 THE MOTOR UNIT

A single α motor neurone which leaves the ventral horn cells of the spinal cord innervates between five and 3000 muscle cells. A motor neurone and the muscle fibres it innervates are together known as a **motor unit**. The ratio of muscle cells to neurones reflects the degree of fine control required, with large muscle groups such as the postural muscles having a much larger ratio than groups such as the muscles of the forearm which bring about finger movement. By contrast, any given muscle cell is innervated by only one neurone.

Action potentials which pass along the axon of a motor nerve cause a Ca^{2+}-dependent release of ACh which binds to the receptors of the motor end-plate of the muscle. This leads to depolarization of the motor end-plate and an action potential in all the muscle fibres of the motor unit. Thus, unlike the many excitatory postsynaptic potentials in a synapse which are required to reach the threshold for firing of the postsynaptic neurone, a muscle action potential is generated for every action potential that reaches the motor end-plate. Increased strength of muscle contraction is brought about by either increasing the number of motor units which are active and/or by increasing the frequency of motor nerve activity. The maximum force of a muscle is produced when the muscle is initially lengthened by about 120%.

The slow twitch (type I) fibres are innervated by small motor neurones which conduct slowly, whereas the fast twitch (type II) fibres are innervated by larger motor neurones with higher conduction velocities. It seems that the motor neurone influences the characteristics of the muscle fibres it innervates since cross-innervation studies (studies where the innervation of one type of motor unit is experimentally altered to the other type of motor unit) have shown that the twitch characteristics can be changed.

3.2.2 SPINAL REFLEX ARCS

The neurones of motor pathways which descend the spinal cord form connections with the α- or γ-motor neurones via interneurones in the spinal cord. These neurones (interneurones and α- and γ-motor neurones) are arranged in circuits with afferent (sensory) neurones from skeletal muscle and thereby form spinal reflex arcs. The afferent fibres from muscle have their origin in the muscle spindles and Golgi tendon organs.

3.2.3 MUSCLE SPINDLES

The muscle spindles are long capsules which are comprised of small muscle fibres either side of an elastic central region. The muscle fibres of the spindles are termed **intrafusal** to distinguish them from the skeletal muscle fibres in which they are located and which are termed **extrafusal**. The intrafusal fibres are attached to tendon or to the endomysium of a fasciculus of extrafusal muscle fibres. They are innervated by the lightly myelinated γ-motor neurones. Firing of these neurones causes the intrafusal fibres to contract, and therefore the elastic central region becomes stretched. The elastic region contains receptors which fire if the degree of stretching is sufficient. Nerve impulses pass along the axons of the afferent neurones arising from the receptors to the spinal cord. The muscle spindles will also fire when the extrafusal fibres are stretched since they lie parallel with them. The result of such afferent activity is reflex stimulation of the α-motor neurones, which causes contraction of the extrafusal fibres. This reflex is **monosynaptic**, that is, there is only one synapse.

There are two kinds of muscle spindle, those in which the sensory portion is in the form of a bag and those in which it is in the form of a chain. The bag intrafusal fibres are innervated by so-called **dynamic** or **type I γ-fibres**. The so-called **static** or **type II γ-fibres** mainly innervate the chain intrafusal fibres and are more numerous than the dynamic fibres. The intrafusal muscle fibres of the bag spindles contract slowly whereas those of the chain spindles develop tension very quickly.

The sensory Ia neurones have their endings in both bag and chain spindles and are characterized by a fast conduction velocity. In contrast the slower conducting type II neurones have their nerve endings predominantly in the chain spindles (table 3.2).

Table 3.2 Distribution of γ (motor) and sensory fibres in muscle

Neurone	Muscle spindle		Golgi tendon organ
	Bag	Chain	
Gamma (motor)			
I (dynamic)	****	–	–
II (static)	*	***	–
Sensory			
Ia (stretch)	**	**	–
Ib	–	–	****
II (length)	*	***	–

The Ia afferent fibres are sensitive to dynamic stretching, firing particularly rapidly during the dynamic stage of muscle fibre lengthening. This provides the CNS with an estimate of the final amount of stretch. The type II afferent fibres are not very sensitive to the dynamic stage of muscle lengthening but do respond to static lengthening of the muscle fibres.

Through their sensitivity to muscle length and their reflex effect on muscle tension the muscle spindles play an important role in ensuring the smooth contraction of skeletal muscle. This is achieved in an integrated manner with the inhibitory effects of the Golgi tendon organs.

3.2.4 GOLGI TENDON ORGANS

The Golgi tendon organs (GTOs) are small sensory organs located in series with the extrafusal muscle fibres next to the tendon. The GTOs respond to changes in muscle tension. As the skeletal muscle fibres contract (or stretch excessively), they squeeze the sensory receptors in the GTOs which, if the threshold tension is attained, cause the afferent Ib sensory neurones to fire. These sensory Ib afferent fibres transmit impulses to interneurones in the spinal cord which in turn synapse with α-motor neurones. They bring about a reflex decrease in the rate of firing of the latter neurones as well as a decrease in the rate of firing of the γ-motor neurones. The GTOs do not require an efferent motor supply and therefore, unlike the muscle spindles, are not directly influenced by the central nervous system.

3.2.5 FINAL COMMON PATH

Motor nerve activity depends on the input from sensory nerves with receptors in the joints, skin and the vestibular apparatus as well as muscle and tendon. These inputs are integrated at the postsynaptic cell and if there are enough excitatory synapses the membrane potential of the postsynaptic cell neurone will be raised sufficiently to initiate an action potential. The motor nerve along which this action potential travels has been termed 'the final common path' (figure 3.2).

3.2.6 RECIPROCAL INHIBITION

In order to achieve joint movement (such as elbow flexion), the agonist muscle, or prime mover, is stimulated to contract while the antagonist muscle is inhibited. This is known as **reciprocal inhibition** or **reciprocal innervation**. In a monosynaptic reflex the Ia fibres from the spindles of the agonist muscle stimulate the α-motor neurones to

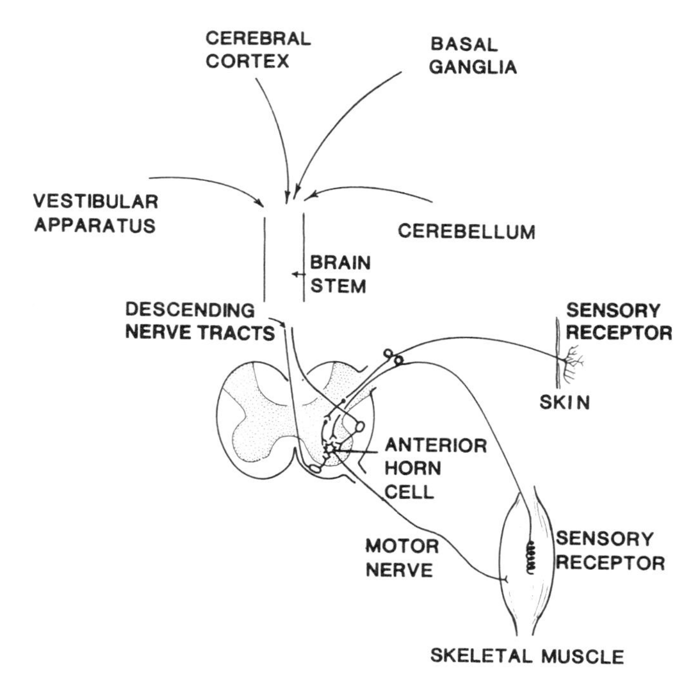

Figure 3.2 Major neuronal influences on the motor nerve.

fire and cause the agonist muscle to contract. These Ia fibres branch into collateral fibres which synapse with inhibitory interneurones in the spinal cord and inhibit the α-motor neurones of the antagonist muscle.

3.3 Control of movement: central mechanisms

3.3.1 INITIATION OF MOVEMENT

The brain may functionally, but not anatomically, be divided into two systems: the limbic system and the sensorimotor system. The limbic system is concerned with emotional behaviour and learning, which it controls through neural connections, via the hypothalamus, between the autonomic nervous system and neuroendocrine mechanisms. The sensorimotor system is concerned with sensation and motor function. Neural integration of the limbic and sensorimotor systems permits volitional skilled movement. The desire of an individual to move occurs in the context of his or her emotional state, previous experience and the environmental setting.

Limbic system		*Sensorimotor system*
Desire to move	→→	Movement

Simultaneous recordings of brain-cell activity and motor-nerve activity indicate that cells of the motor cortex discharge before the muscles contract. In addition, since there are anatomical connections between the motor cortex and skeletal muscles, it seems reasonable to think that movement is initiated by cells in the motor cortex. However, nerve cells in the cerebellum and part of the basal ganglia also fire prior to muscle contraction. Both of these regions receive nerve fibres from the cerebral motor cortex and send fibres via the thalami back to the cortex. Through these anatomical connections the cerebellum and basal ganglia are thought to play an important role in the coordination of movement. It is now known that all three areas are involved in the control of movement, and they are described in more detail below.

3.3.2 MOTOR CORTEX

The pyramidal cells of the cerebral motor cortex, which lie anterior to the central sulcus of the cortex, are arranged in columns. These columns of cells send descending fibres to the spinal interneurones which synapse with α- and γ-motor neurones and leave the spinal cord via the ventral horn cells to different muscle groups. The various columns have control over specific muscles. The specificity of these columns has led to the development of a figurine of the body, which appears schematically upside down in the cerebral motor cortex. This 'body map' is characterized by having a large face, tongue and hands relative to the rest of the body, and this represents the large number of nerve cells that have axons descending the spinal cord with motor outflow to these regions. Pathways from the motor cortex not only directly descend the spinal cord via the pyramidal tract but also branch to brainstem nuclei and the cerebellum.

3.3.3 CEREBELLUM

The cerebellum receives sensory input peripherally from skin, joints, skeletal muscle, tendons, the vestibular apparatus and the eyes as well as centrally from other neurones in the brain. These afferent neurones pass either directly to the cerebellum via the spinocerebellar tracts or indirectly via the thalamus and cerebral cortex. The descending neurones pass via the thalamus, red nucleus and nuclei of the brainstem. They reach the motor neurones either through the pyramidal tract or through the rubrospinal (from the red nucleus) and vestibulospinal tracts.

3.3.4 BASAL GANGLIA

The basal ganglia form the third major morphological brain division which makes up the central motor control system. This division regulates stereotyped activity and initiates complex volitional movements. The basal ganglia act as relay stations which link the cerebral cortex with nuclei in the thalamus.

3.3.5 OTHER CENTRAL CONTROL STRUCTURES

Limbic structures also evoke motor activity. This motor activity is related to feeding (such as chewing and swallowing), behavioural activity (such as grooming) and self-preservation (such as growling and clawing).

There is also evidence of a mechanism for automatically generating movement from the spinal cord in the absence of supraspinal inputs. This evidence comes from studies of spinal animals (animals in which the spinal cord has been separated from the brain by a lesion at the first cervical vertebra. These animals are able to spontaneously stand, produce stepping movements and scratch. However, spinally generated movement patterns do not display the fine control that animals (including man) have over volitional movement. For this, the integration of spinal and supraspinal mechanisms is essential.

3.4 Autonomic function during exercise

Physical activity leads to potentially devastating disturbances of the body's internal environment. Many physiological mechanisms are involved in bringing about physiological responses appropriate to maintaining homeostasis. In this regard the autonomic nervous system plays a key role. Some of these autonomic adjustments (for example an increase in heart rate and skeletal muscle vasodilatation) actually occur in the anticipatory stage of exercise.

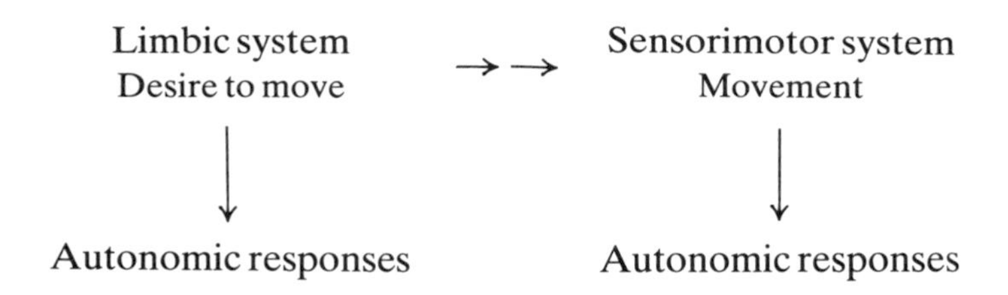

During exercise a variety of neural signals are received by cells in the spinal cord, brainstem and hypothalamus. The source of this information is widespread, including the central motor system, the limbic system, peripheral receptors and blood-borne variables (for example temperature, pO_2 and pCO_2). Cells of the spinal cord, brainstem and hypothalamus are critical for integrating this information and ensuring the most appropriate spinal output via autonomic neurones (figure 3.3). The peripheral organs which are innervated by autonomic nerves, and the homeostatic adjustments made as a result of autonomic activity and increase in adrenal medullary secretion of catecholamines during exercise, are illustrated in figure 3.4.

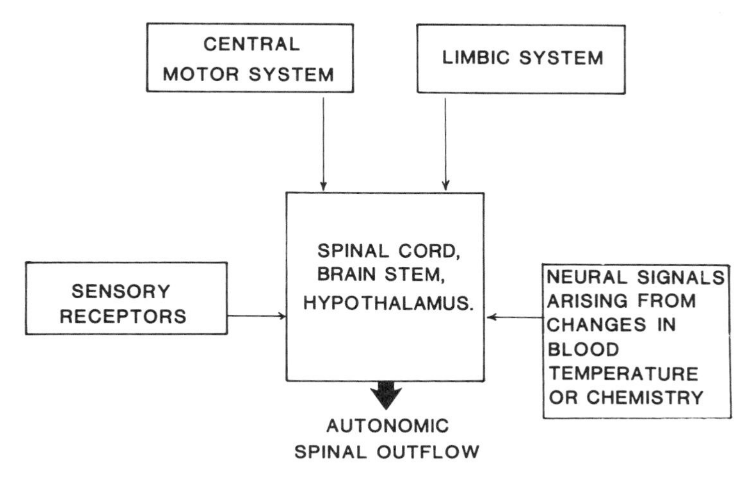

Figure 3.3 Major neuronal influences on autonomic spinal outflow.

These autonomic responses can be seen to fall broadly into three categories:

1. metabolic (liver and muscle glycogenolysis, lipolysis and pancreatic responses)
2. cardiovascular (inotropic and chronotropic effects on the heart, decrease in liver and gastrointestinal tract blood flow)
3. thermoregulatory (sweating).

These three categories of autonomic response to exercise will be discussed in the following three chapters.

Autonomic activity may be associated with a discrete response in one part of the body as seen, for example, during heat stress. Heat stress is associated with sweating and an increase in skin blood flow, which occur in the absence of more widespread sympathetic responses. In contrast, widespread sympathetic responses are seen during vigorous sustained exercise. The widespread sympathetic responses constitute the 'fight or flight' syndrome originally described by Cannon (1915). Since the body's physiological systems are linked such that a change in the functioning of one system potentially affects another, it becomes difficult to consider the effect of exercise on each of the physiological systems in isolation from each other.

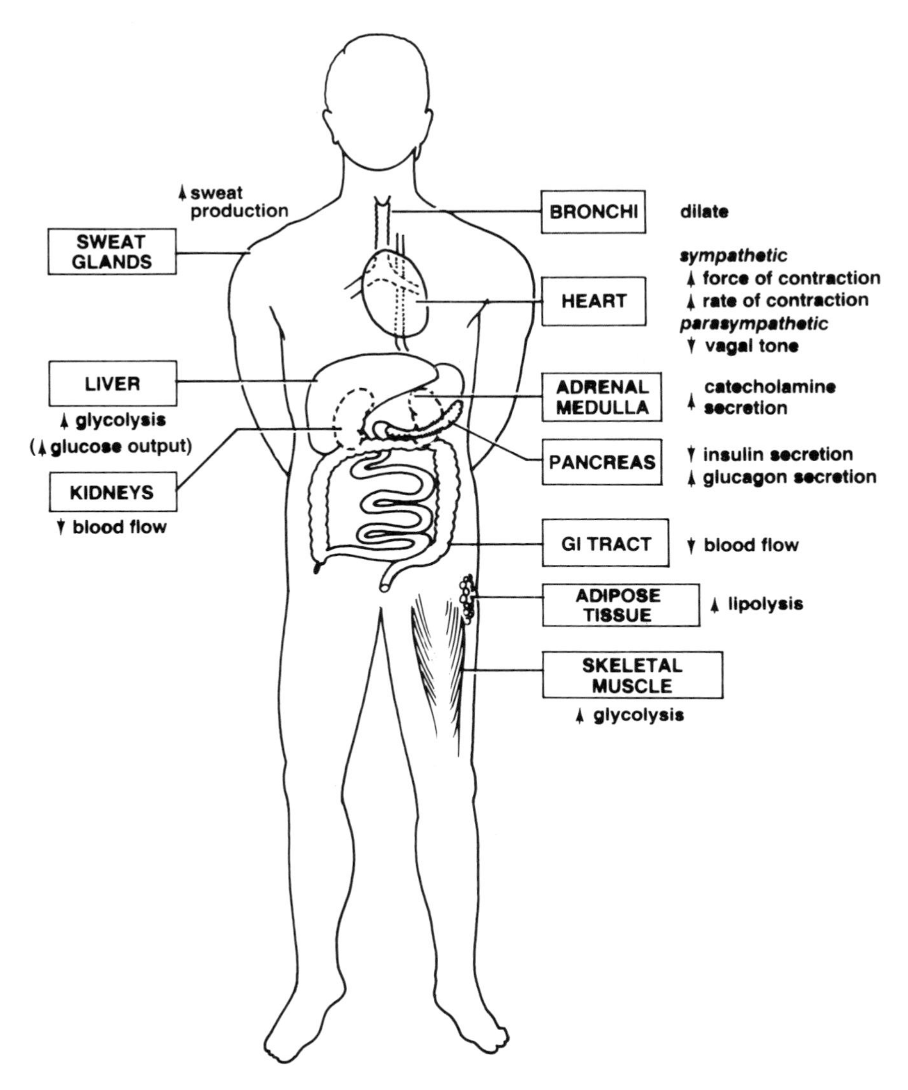

Figure 3.4 Major autonomic responses during exercise.

Thus the autonomic nervous system may elicit discrete physiological adjustments to control homeostasis or it may produce a widespread 'stress' response. In the case of exercise, the extent of sympathoadrenal involvement in mediating physiological responses to exercise depends on the duration and intensity of work, the type of work, the climatic conditions and the characteristics of the individual (such as age, sex, fitness, health and disease).

References and further reading

Brooke, M.H. and Kaiser, K.K. (1970) Muscle fibre types: How many and what kind? *Arch. Neurol.*, **23**, 369–379.

Brooks, V.B. (1986) *The Neural Basis of Motor Control.* Oxford University Press, Oxford.

Buchthal, F. and Schmalbruch, H. (1980) Motor unit of mammalian muscle. *Physiol. Rev.*, **60**, 90–142.

Burke, D. (1980) Muscle spindle function during movement. *Trends Neurosci.*, **3**, 251–253.

Cannon, W.B. (1915) *Bodily changes in Pain, Hunger, Fear and Rage.* Harper and Row, New York.

Costill, D., Daniels, J., Evans, W., Fink, W., Krahenbuhl, G. and Saltin, B. (1976) Skeletal muscle enzymes and fiber composition in male and female track athletes. *J. Appl. Physiol.*, **40**, 149–154.

Evarts, E.V. (1973) Brain mechanisms in movement. *Sci. Am.*, **241**, 164–179.

Gaesser, G.A. and Brooks, G.A. (1975) Muscular efficiency during steady-rate exercise: effects of speed and work rate. *J. Appl. Physiol.*, **38**, 1132–1139.

McGeer, P.L. and McGeer, E.G. (1980) The control of movement by the brain. *Trends Neurosci.*, **3**, III-IV.

Merton, P. (1972) How we control the contraction of our muscles. *Sci. Am.*, **226**, 30–37.

Pollack, G.B. (1983) The cross-bridge theory. *Physiol. Rev.*, **63**, 1049–1113.

Tracey, D.J. (1980) Joint receptors and the control of movement. *Trends Neurosci.*, **3**, 253–255.

Chapter 4

Energy metabolism

There are only three types of fuel available to the cells: glucose, fatty acids and amino acids. These fuels are stored principally in skeletal muscle, the liver and adipose tissue. The cells of the nervous system have a crucial requirement for glucose but in times of shortage, such as five or six days of starvation, they are able to use ketone bodies formed from fatty acids. There is also evidence of an increase in ketone utilization during endurance exercise and following endurance training.

Fuel metabolism is controlled by the local increase in Ca^{2+} ions as well as by the autonomic nervous system and by hormones. In turn the metabolic pathways are regulated by key enzymes which limit or enhance the rate of the sequence of reactions in the pathway. The major hormones involved in controlling metabolism are adrenaline, insulin, glucagon, cortisol, growth hormone and thyroxine and, in sufficiently high concentrations, noradrenaline (Chapter 2). The metabolic effects of the catecholamines may occur as a direct result of catecholamine action or as an indirect consequence of their effect on other hormones (figure 4.1). The metabolic effects of only the catecholamines, insulin and glucagon will be considered in this chapter. First, however, an overview of the cellular events in exercise metabolism is presented.

4.1 Energy for muscular contraction

Energy for muscular contraction is derived from adenosine triphosphate (ATP). This crosses cell membranes poorly and is not transported in the blood. It must be generated in the tissues, such as skeletal muscle, where it is used. This generation of ATP is essential since ATP is not stored in great enough quantities to support muscle

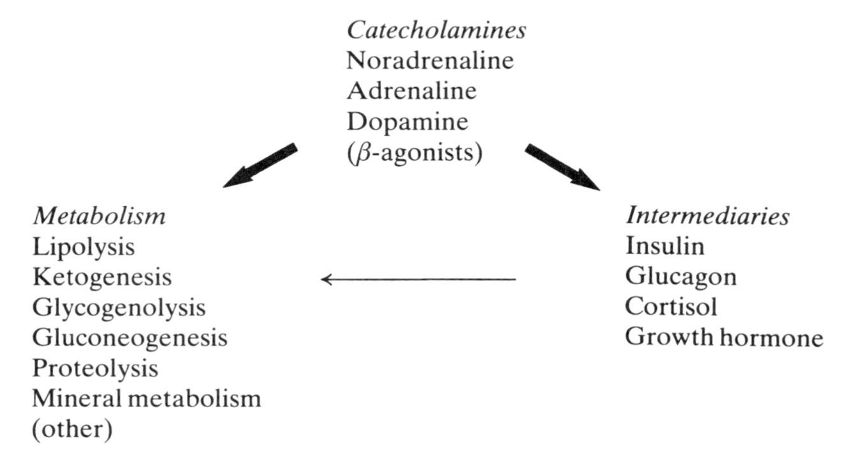

Figure 4.1 Effects of catecholamines on metabolism. Reprinted by permission from *Clinical Science*, Volume 68, page 613. Copyright © 1985, The Biochemical Society, London.

contraction for more than a few seconds. The blood-borne fuels from which ATP is derived during exercise are glucose and free fatty acids (FFA). Glucose in blood is derived mainly from liver glycogen although some is also derived from lactate. In addition muscle has its own store of glycogen to provide ATP necessary for muscular contraction. Free fatty acids are derived from triacylglycerol stored in adipose tissue. However, there is also evidence that intramuscular triacylglycerol is used to derive ATP for exercise. The balance between these fuel supplies depends on the individual's endowment of slow twitch and fast twitch fibres and the type of exercise being performed. A good-class sprinter is likely to have more than 70% fast twitch fibres which have a very poor blood supply but a good capacity to break down glycogen in the absence of oxygen. Such an athlete must therefore rely on endogenous fuel, that is, muscle glycogen. Although muscle glycogen stores are limited, this is not a problem for an athlete performing short duration work. Since fast twitch muscle is endowed with a comparatively good supply of both glycogen and glycolytic enzymes compared with slow twitch muscle, it is possible to meet the energy requirements of exercise through anaerobic glycolysis. In contrast a good-class marathon runner is likely to have more than 70% slow twitch fibres which have a rich blood supply and can therefore receive transported fuel (glucose and FFA) from exogenous depots. Once delivered to the muscle, the slow twitch fibres, unlike the fast twitch fibres, are suited to using glucose and FFA due to the greater aerobic capacity which is a result of their rich supply of myoglobin and mitochondria.

In terms of weight the major fuel store in man is lipid. Lipid can be stored easily since it has a high energy density with little water content and is therefore ideally suited for long-term survival needs. However, since FFA uptake into cells is limited by the concentration in blood, it cannot supply the energy demands of the exercising individual alone. The blood level is restrained to a maximum of only around 2 mmol $litre^{-1}$, with higher levels being toxic to the cells. Free fatty acids can never provide more than about half of the ATP production during exercise. However, this contribution is critical during endurance events of more than 1–2 h. It is essential to maintain blood glucose since nerve cells have a priority for glucose. During exercise prolonged over several hours, liver glycogen depots are insufficient to maintain blood glucose. There is an increased reliance on gluconeogenesis (section 4.2.2) to maintain blood glucose. In addition, at these times the use of FFA derived from fat metabolism becomes important. An individual who is in fluid and electrolyte balance does not become exhausted until muscle glycogen is depleted, and this, in effect, can be delayed by the use of FFA, providing the rate of work is sufficiently low (figure 4.2).

4.1.1 MOLECULAR ENERGY

Since energy is never created or lost, the energy for muscle contraction must be derived from redistributing energy in the body. At the cell level this is accomplished by breaking or rearranging the chemical bonds which link the atoms in molecules.

Carbohydrates

Carbohydrates are formed from carbon, hydrogen and oxygen atoms in the general form of $C_n(H_2O)_n$

$$\left(\begin{array}{c} \mathrm{H} \\ | \\ -\mathrm{C}- \\ | \\ \mathrm{OH} \end{array} \right)_n$$

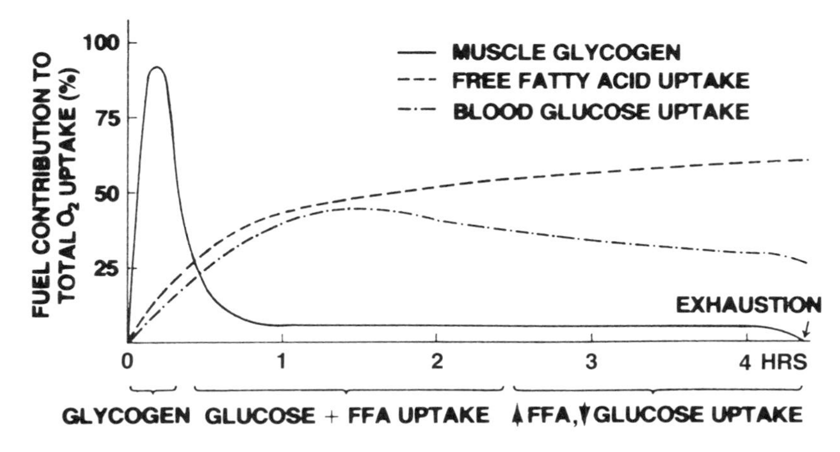

Figure 4.2 Contribution of metabolic fuels during prolonged exercise. Reproduced with permission from Felig and Koivisto (1979).

Glucose takes the form of $C_6H_{12}O_6$. In glycogen, carbons 1 and 4 are linked through oxygen with the loss of water (figure 4.3). The hydroxyl groups interact with hydrogen (from H_2O) in the cell ordering water around the glycogen molecule. Glycogen is a branched molecule with branches where carbon 6 of one residue is linked to carbon 1 of another. This branching is of functional significance because the enzyme glycogen phosphorylase, which breaks down glycogen when it is required to supply ATP, acts only on the terminal branches of the molecule. With glycogen having a branched structure, glycogen phosphorylase has many sites to work on, thereby facilitating the supply of ATP.

Lipids

The majority of atoms in a lipid molecule are carbon and hydrogen. The general structure of the fatty acids is

$$HO-\overset{\overset{\displaystyle O}{\|}}{C}-CH_2(CH_2)_n-CH_3$$

Fatty acids are stored in adipose tissue in the form of triacylglycerol which comprises three molecules of fatty acid and one of glycerol

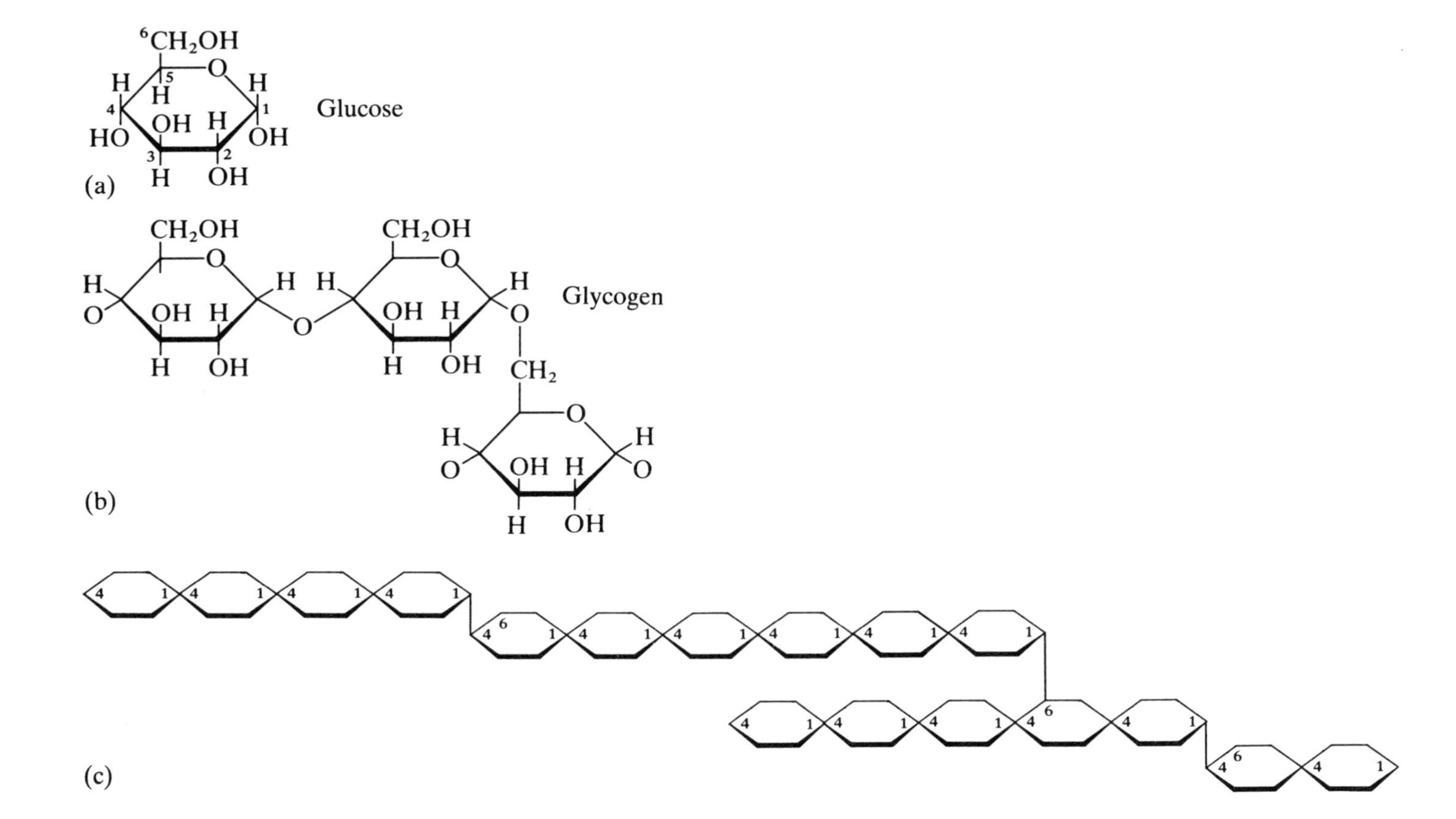

Figure 4.3 Glycogen. (a) Glucose; (b) glycogen; (c) branched structure of glycogen.

(figure 4.4). The fatty acids are 1–24 carbon atoms long but they are more commonly 16–20 carbon atoms long.

4.1.2 BOND ENERGY

The energy that is stored in the bonds of molecules can be determined by burning the substance and measuring the heat that results. By burning compounds with known molecular structures it can be shown that each carbon–carbon or carbon–hydrogen bond is worth 255 kJ. However, a nitrogen–carbon or nitrogen–hydrogen bond in protein is worth only 82 kJ. Thus the energy in carbohydrates, fats and protein can be estimated from looking at their molecular structures. The C–C, C–H, N–C and N–H bonds release the energy when the atoms are oxidized. Thus, there is no energy in carbon dioxide (O═C═O) because the carbon atoms are oxidized or in water (H–O–H) because the hydrogen atoms are oxidized.

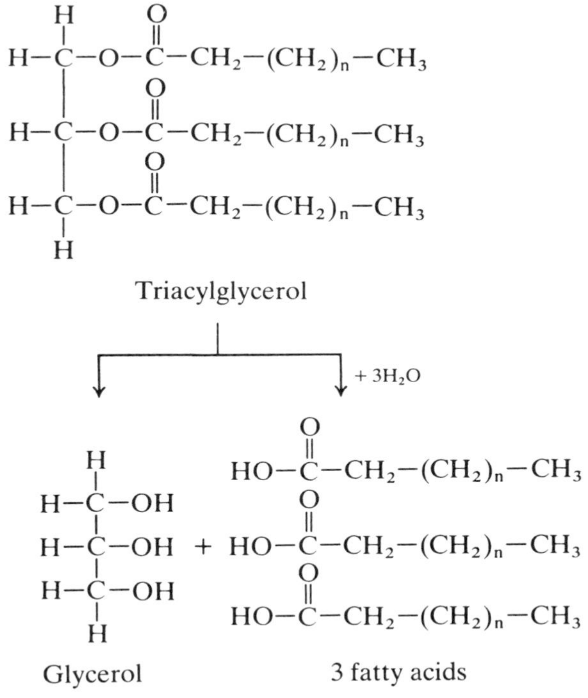

$$
\begin{array}{l}
\mathrm{H{-}CH{-}O{-}C({=}O){-}CH_2{-}(CH_2)_n{-}CH_3} \\
\mathrm{H{-}C{-}O{-}C({=}O){-}CH_2{-}(CH_2)_n{-}CH_3} \\
\mathrm{H{-}CH{-}O{-}C({=}O){-}CH_2{-}(CH_2)_n{-}CH_3}
\end{array}
$$

Triacylglycerol

$$\downarrow + 3H_2O$$

$$
\begin{array}{lcl}
\mathrm{H{-}CH{-}OH} & & \mathrm{HO{-}C({=}O){-}CH_2{-}(CH_2)_n{-}CH_3} \\
\mathrm{H{-}C{-}OH} & + & \mathrm{HO{-}C({=}O){-}CH_2{-}(CH_2)_n{-}CH_3} \\
\mathrm{H{-}CH{-}OH} & & \mathrm{HO{-}C({=}O){-}CH_2{-}(CH_2)_n{-}CH_3}
\end{array}
$$

Glycerol 3 fatty acids

Figure 4.4 Triacylglycerol, glycerol and three free fatty acids.

Typically, three (but sometimes two) molecules of ATP are made available for every C–C or C–H bond that is broken. For this reason the energy yield from breaking down a molecule of fat with its low number of oxygen atoms far exceeds that from breaking down a molecule of glucose with its comparatively high number of oxygen atoms. When expressed per gram of fuel this difference is further magnified. Since carbohydrate is much heavier than lipid due to the oxygen atoms in its structure, there are less molecules per gram of carbohydrate than of lipid. Thus, in addition to the nature of the molecule, chemical energy also depends on the number of molecules available.

In the cell the energy yield for glycogen is not realized purely on a weight basis since it is stored with water. Any weight loss of glycogen includes water loss. Around 60% of the weight of glycogen is actually water.

Chemical reactions

Chemical energy is a form of potential energy stored in the structure of the molecules. Molecules are in a constant state of motion. A reaction occurs when the electrons in the outer shells of two molecules interact. This interaction requires energy and produces energy due to the breaking of atomic bonds. An endergonic reaction is one where there is a net input of energy while an exergonic reaction is one where there is a net output of energy. In teleological terms the latter is more favourable when the required outcome is the availability of energy.

If an exergonic reaction takes place in a test tube, it will be accompanied by the generation of heat. Calorimetry measurements indicate that 3060 kJ are derived per mole of glucose completely broken down (6×10^{23} molecules $\times$ 12 C–C/C–H bonds each worth 255 kJ). Although heat is a form of energy, it is not a form that can be used by the cells. The cells are able to use the energy released from hydrocarbon bonds in the synthesis of other molecules available in the cell, predominantly ATP. Thus, in the cell, glucose oxidation is coupled to ATP formation.

$$C_6H_{12}O_6 + 6O_2 \xrightarrow[\text{ADP} \rightarrow \text{ATP}]{} 6H_2O + 6CO_2$$

ATP has a crucial role in transferring the energy released in exergonic reactions to endergonic reactions, with 30.5 kJ being transferred from the hydrolysis of the terminal phosphate in ATP and 58.1 kJ from the hydrolysis of both high-energy phosphates. For every mole of glucose completely broken down, 38 mol of ATP are generated (39 for glycogen), thus 2208 kJ of energy are stored within the cell as ATP (that is, about 72% of the total energy contained in the glucose). A similar value is found for fatty acids.

4.2 Metabolic pathways

4.2.1 GLYCOLYSIS

Glycolysis means the splitting of glucose. The glycolytic pathway can be divided into three stages:

1. the transport of glucose into the cell and the formation of glucose-6-phosphate
2. the conversion of glucose-6-phosphate to two triose phosphates
3. the conversion of the triose phosphates into pyruvate.

Subsequently, pyruvate is either converted into lactate (anaerobic glycolysis) or it is oxidized in the mitochondria to acetyl coA which then enters the **tricarboxylic acid (TCA) cycle** of reactions.

The Embden–Meyerhof pathway

The extramitochondrial stages of glucose breakdown are often referred to as the **Embden–Meyerhof pathway** (figure 4.5), named after two German biochemists who made a large contribution to this field of knowledge. Glycolysis begins with glucose transport across the cell membrane down a concentration gradient. Glucose is unable to pass across the membrane freely and in skeletal muscle the hormone insulin stimulates the activity of a protein molecule which facilitates glucose transport. This process of facilitated diffusion does not require energy since the concentration gradient is favourable. During exercise some insulin is required but an independent mechanism is involved in glucose transport across the muscle cell membrane.

As soon as glucose enters the cell it is phosphorylated (gains a phosphate group) (reaction 1, figure 4.5) to become glucose-6-phosphate (G-6-P) which is an irreversible reaction. For this reason,

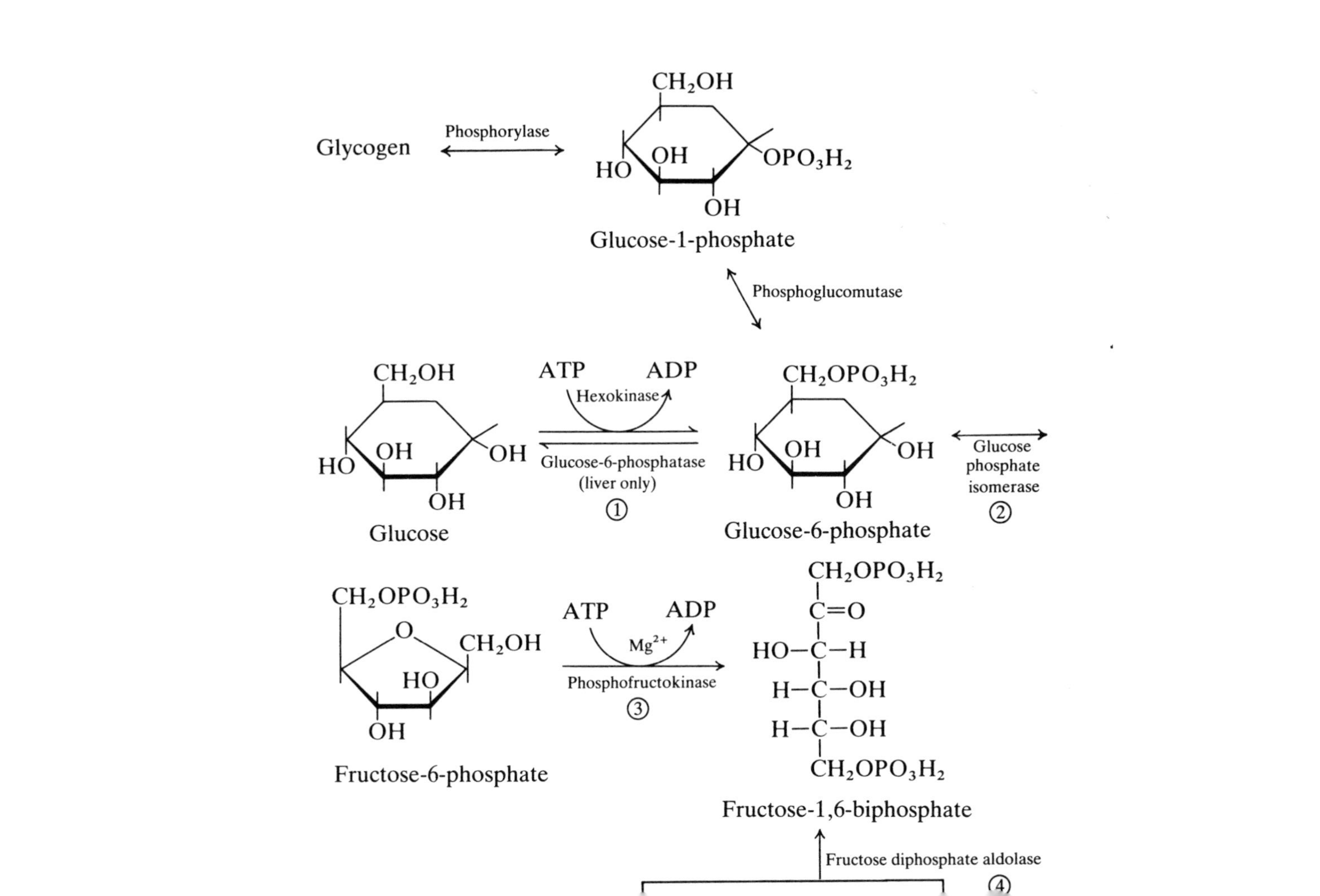
Glycogen
Phosphorylase
Glucose-1-phosphate
Phosphoglucomutase
Glucose
ATP
ADP
Hexokinase
Glucose-6-phosphatase
(liver only)
①
Glucose-6-phosphate
Glucose
phosphate
isomerase
②
Fructose-6-phosphate
Mg2+
Phosphofructokinase
③
Fructose-1,6-biphosphate
Fructose diphosphate aldolase
④

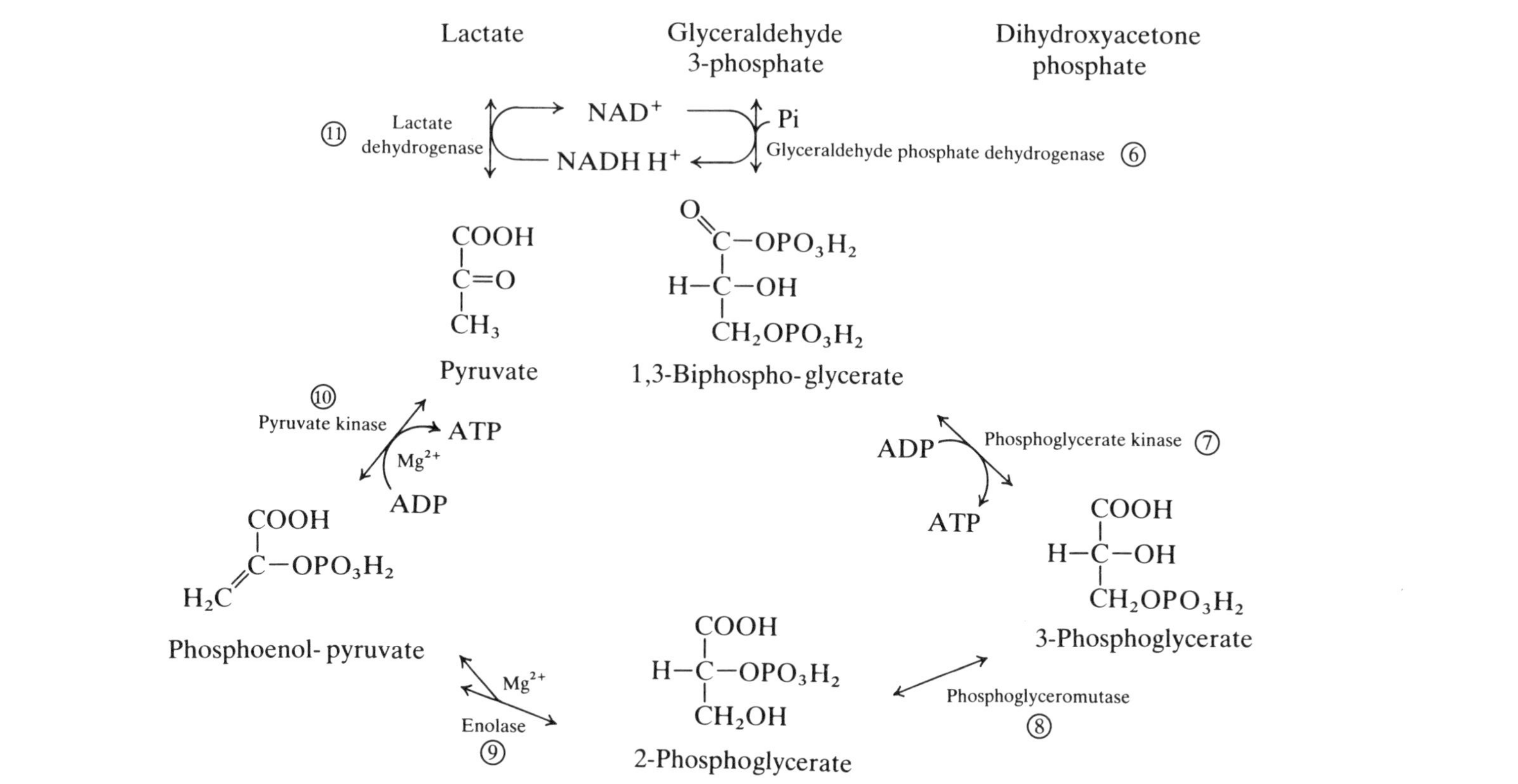

Figure 4.5 Glycolysis.

and because G-6-P cannot pass easily across the cell membrane, glucose becomes trapped in the cell as phosphorylated glucose. This lowers the intracellular glucose concentration, which has the effect of raising the concentration gradient between the outside and inside of the cell, favouring further glucose entry into the cell down the concentration gradient.

The third reaction of glycolysis is also irreversible and is therefore a second major control point in the pathway (discussed later in this section). It can be seen from figure 4.5 that, in the first and third reactions which involve phosphorylation, ATP is used to supply the phosphate group. Thus, two molecules of ATP have been used in the pathway, which ultimately generates ATP.

The fourth reaction involves the splitting of the six-carbon molecule fructose-1,6-biphosphate into two three-carbon molecules which are known as triose phosphates. The two molecules are glyceraldehyde-3-phosphate and dihydroxyacetone phosphate. Only glyceraldehyde-3-phosphate is directly converted into pyruvate. As the concentration of glyceraldehyde-3-phosphate decreases (with the production of pyruvate), dihydroxyacetone phosphate is converted into glyceraldehyde-3-phosphate (reaction 5). Thus from now on in glycolysis two three-carbon molecules of each compound are made per six-carbon molecule of glucose that enters the pathway.

The immediate fate of glyceraldehyde-3-phosphate is its conversion into 1,3-biphosphoglycerate. The biochemistry of this sixth reaction is complex. The essential points are that the molecule becomes phosphorylated (not this time at the expense of ATP) and is oxidized (loses electrons). **Nicotinamide adenine dinucleotide (NAD^+)** is reduced during the oxidation of glyceraldehyde-3-phosphate to become NADH + H^+. In this way NAD^+ acts as a chemical carrier for electrons and because of this function NAD^+ is an important co-enzyme in oxidation-reduction reactions. Another co-enzyme which functions in this way is **flavin adenine dinucleotide (FAD)** (section 4.2.3).

ATP for muscular contraction is not formed until the seventh reaction in which a phosphate group from 1,3-biphosphoglycerate is transferred to ADP to form ATP. This way of forming ATP (by transferring a phosphate group from an intermediate compound in the pathway to ADP to form ATP) is called **substrate level phosphorylation**. Since for every six-carbon glucose molecule two three-carbon molecules of 3-phosphoglycerate are formed, two molecules of ATP are formed at this stage of glycolysis. These two ATP molecules represent the net profit of ATP in anaerobic glycolysis.

However, if the pathway is to run again it is necessary to replace the ATP used in the first and third reactions of the pathway. These are derived from a substrate level phosphorylation in the tenth reaction in which a phosphate group from phosphoenolpyruvate is transferred to ADP to form ATP (two molecules of ATP per glucose molecule).

However, the pathway also requires a continual supply of NAD^+ in the sixth reaction. During maximal muscular effort the supply of NAD^+ in the muscle would last only about three seconds. There are two possibilities for regenerating NAD^+. Either an oxygen atom can be used to regenerate NAD^+ from NAD + H^+:

$$NADH + H^+ + O \rightarrow NAD^+ + H_2O$$

or lactate is produced. The conversion of pyruvate to lactate (reaction 11) involves the transfer of the two electrons from NADH + H^+ to pyruvate to form lactate. Thus the NAD^+ pool is maintained. It is important to recognize that there are the same number of C–C and C–H bonds in two molecules of lactate as there are in one molecule of glucose. The reason why energy can be derived from glucose without the oxidation of these carbon and hydrogen atoms lies in the ability of the cell to generate ATP from substrate-level phosphorylation. Thus the production of lactate does not require oxygen, hence the term **anaerobic glycolysis**. Anaerobic glycolysis is of particular importance in the white skeletal muscle fibres which have a relatively poor blood (and therefore oxygen) supply compared with the red fibres. However, it cannot be assumed that lactate is produced *only* if there is insufficient oxygen since it is produced during exercise even when the muscle pO_2 is high. Experiments using intravenous infusions of catecholamines in *in situ* dog skeletal muscle suggest that adrenaline (more than noradrenaline) mediates an increase in lactate output from skeletal muscle (and other tissues) whilst venous oxygen concentration and partial pressure are maintained.

From the muscle, lactate is transported in the blood to the liver, heart and red muscle. During and following exercise the liver is the main organ that disposes of lactate. In the liver, lactate is a precursor used in gluconeogenesis.

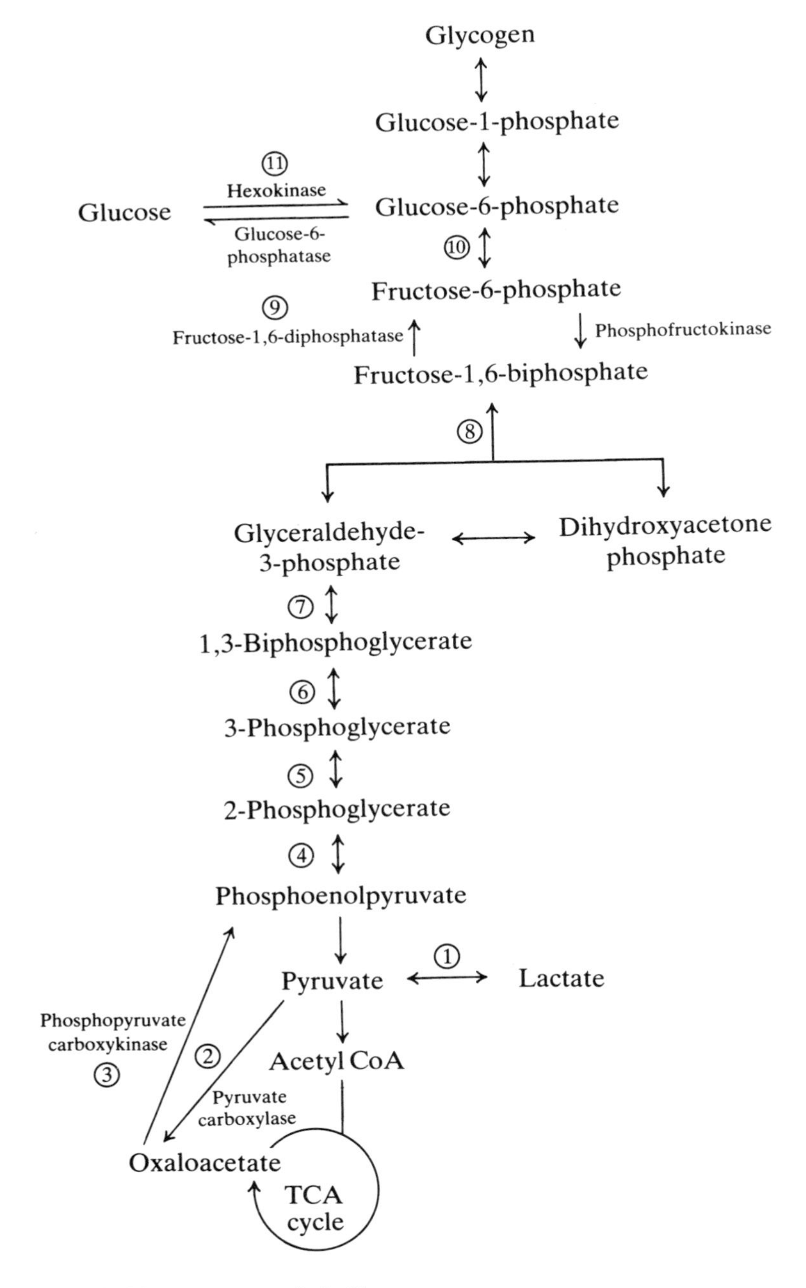

Figure 4.6 Gluconeogenesis in liver.

4.2.2 GLUCONEOGENESIS

Gluconeogenesis is the reversal of glycolysis. Most of the enzymes used in the pathway for gluconeogenesis are the same as those used in glycolysis. However, two reactions are irreversible if glycogen is produced and three reactions are irreversible if glucose is produced. These reactions provide an important mechanism for regulating the net breakdown or production of glucose depending on the metabolic status and energy requirements of the cell (figure 4.6).

Oxidation of pyruvate

The oxidation of pyruvate takes place in the mitochondria. Pyruvate enters the mitochondria where it reacts with coenzyme A and is decarboxylated (loses CO_2) to form acetyl coA. Coenzyme A is a large molecule with a small reactive site in the form of a sulphydryl group (SH). Acetyl coA then enters the TCA cycle, which is the final common pathway for the aerobic breakdown of carbohydrate, lipids and proteins.

4.2.3 THE TCA CYCLE (FIGURE 4.7)

In the TCA cycle there is only one substrate-level phosphorylation and this occurs in the fifth reaction when guanine triphosphate (GTP) is formed which is readily converted to ATP:

$$\text{GTP} + \text{ADP} \leftrightarrow \text{ATP} + \text{GDP}$$

Overall four pairs of hydrogen atoms (or rather their electron equivalents see table 4.1) are formed per cycle of the pathway, $3 \times \text{NADH} + \text{H}^+$ and $1 \times \text{FADH}_2$. However, it should be remembered that the TCA cycle runs twice per six-carbon glucose molecule that enters the glycolytic pathway since each generates two three-carbon molecules of acetyl coA. The pairs of electrons thus generated are subsequently passed along the electron transport chain located in the inner membrane of the mitochondria for the generation of ATP by a process called oxidative phosphorylation.

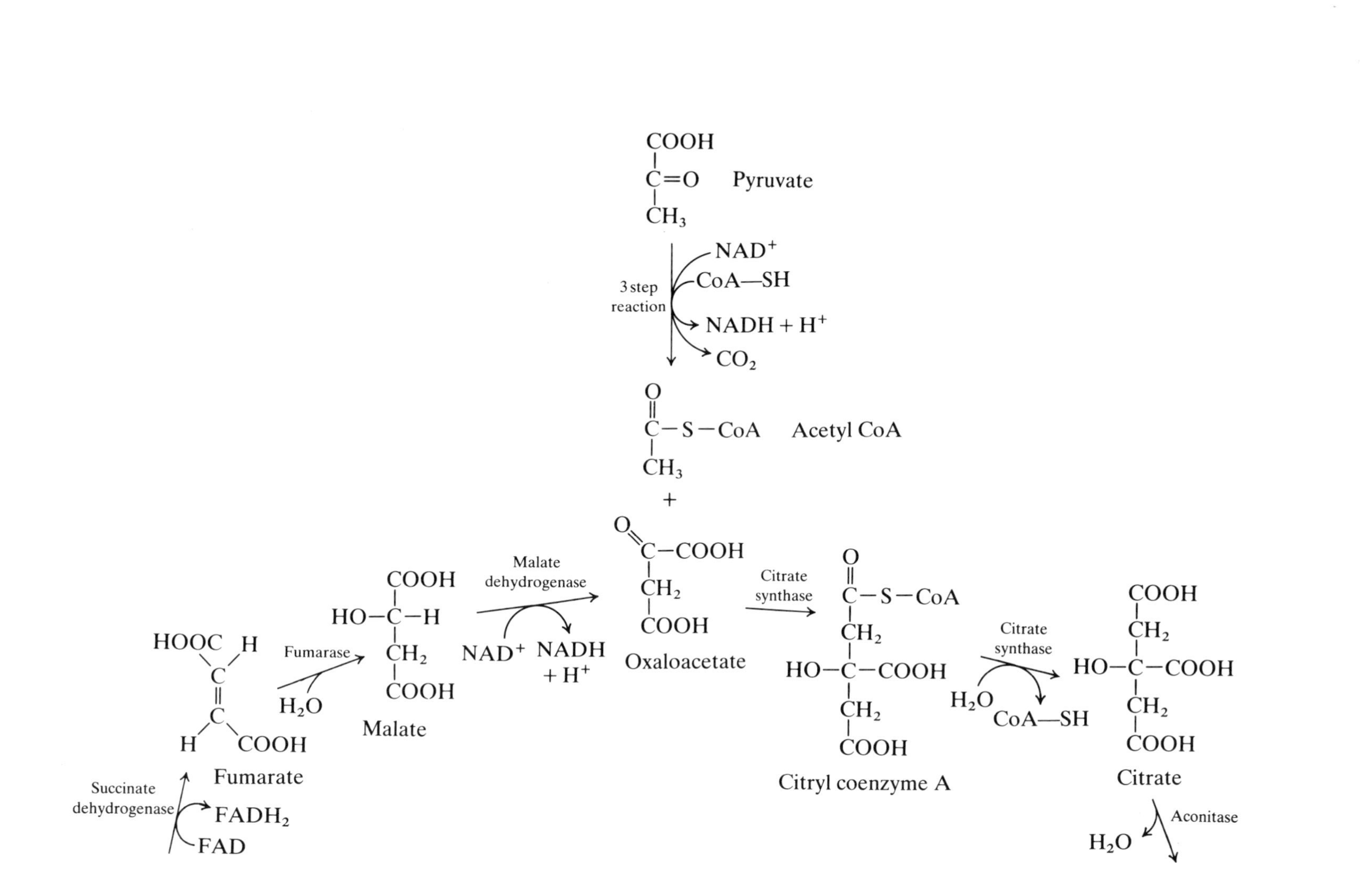
Pyruvate
3 step reaction
NAD⁺
CoA—SH
NADH + H⁺
CO₂
Acetyl CoA
Oxaloacetate
Citrate synthase
Citryl coenzyme A
Citrate synthase
H₂O
CoA—SH
Citrate
Aconitase
H₂O
Malate dehydrogenase
NAD⁺
NADH + H⁺
Malate
Fumarase
H₂O
Fumarate
Succinate dehydrogenase
FADH₂
FAD

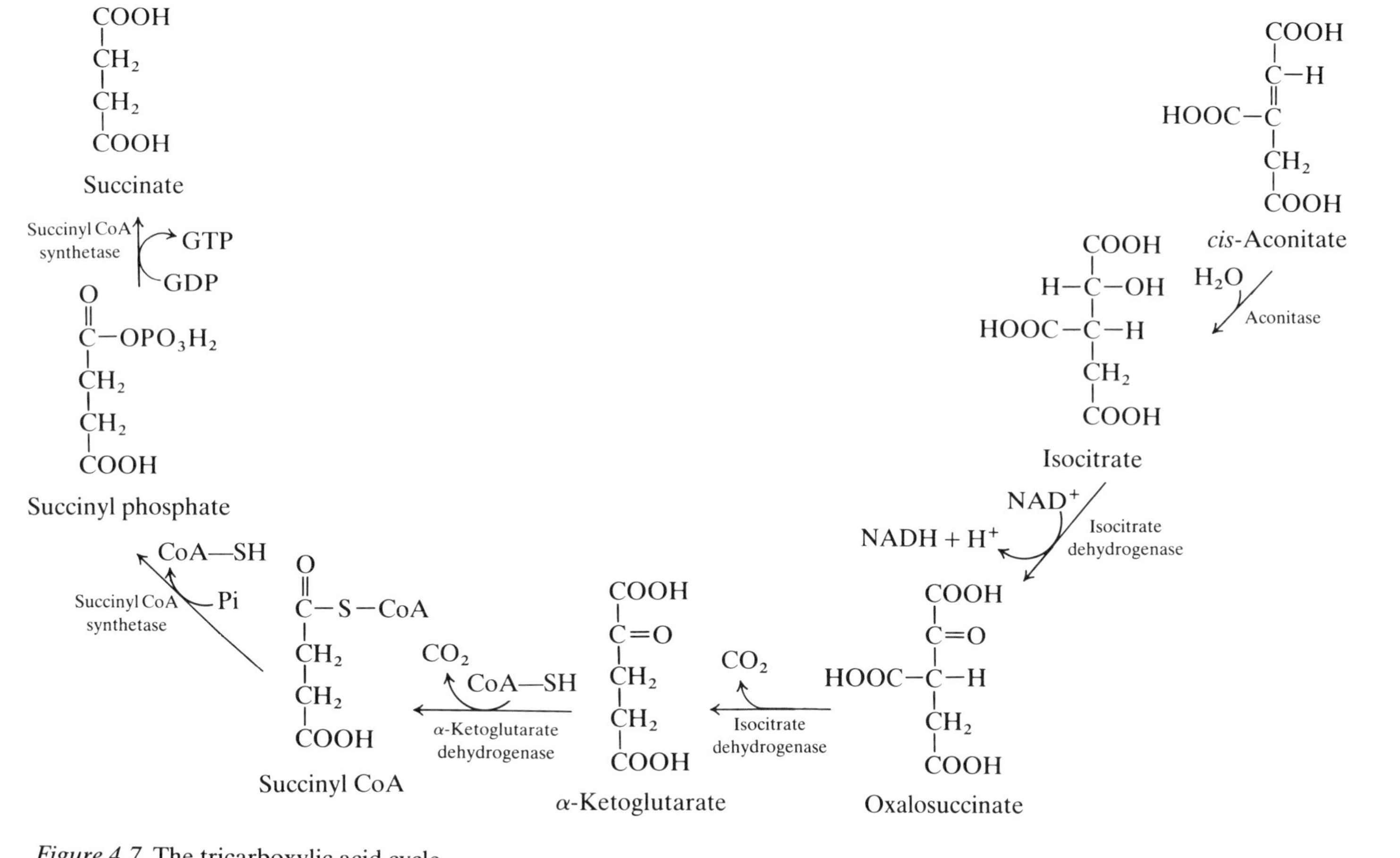

Figure 4.7 The tricarboxylic acid cycle.

Table 4.1 Oxidation-reduction, with two examples of half reactions

Oxidation		Reduction
Removal of electrons		Addition of electrons
$NAD^+ + 2H^+ + 2e^-$	$\longrightarrow$	$NADH + H^+$
Pyruvate $+ 2H^+ + 2e^-$	$\longrightarrow$	Lactate

Oxidative phosphorylation

Oxidative phosphorylation does three things:

1. It produces innocuous end-products;
2. It drives ATP production;
3. It restores the supply of NAD^+ and FAD.

However, when reduced these latter two enzymes NAD^+ and FAD cannot pass their electrons directly to oxygen. Instead, their electrons are transferred between carrier molecules. Such molecules are **redox couples**, that is, compounds which are first reduced and then oxidized or vice versa. These carriers are located in the mitochondrial inner membrane. They are aligned so that there is a gradual drop in the standard electrode potential between NADH and oxygen. Oxygen is at a high standard electrode potential, which results in electrons from other molecules having a tendency to flow towards it. The electrode potential drop is divided into three sections (table 4.2).

As the electrons are passed between redox couples, hydrogen ions are pumped across the inner mitochondrial membrane. The potential energy in this electrochemical gradient is converted to potential energy in the form of ATP by the re-entry of the hydrogen ions to a

Table 4.2 Standard electrode potential (E^{θ}) of some redox couples of biological interest in the mitochondrial inner membrane

Couple	E^{θ} (mV)
$NADH + H^+/NAD^+$	−320
Ubiquinol (QH_2)/ubiquinone (Q)	−90
Reduced cytochrome *c*/cytichrome *c*	+230
H_2O/O_2	+820

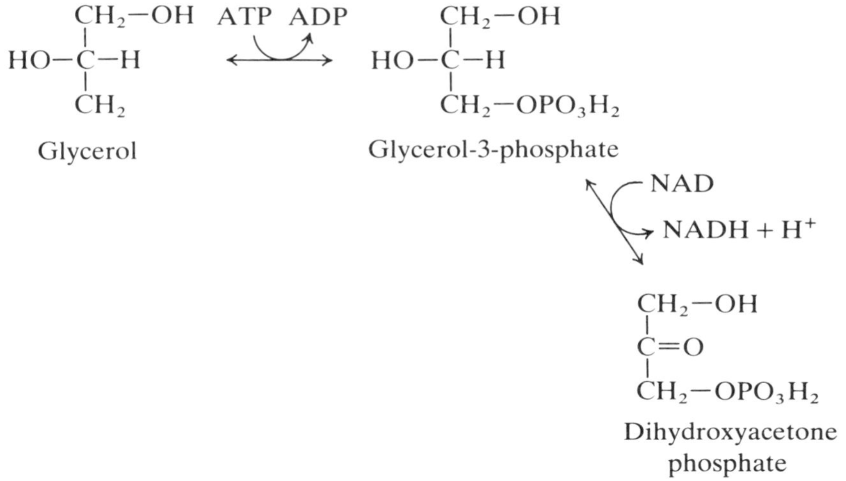

$$\begin{array}{c} CH_2-OH \\ | \\ HO-C-H \\ | \\ CH_2 \end{array} \xrightleftharpoons{ATP \quad ADP} \begin{array}{c} CH_2-OH \\ | \\ HO-C-H \\ | \\ CH_2-OPO_3H_2 \end{array}$$

Glycerol — Glycerol-3-phosphate

$$\downarrow NAD \rightarrow NADH + H^+$$

$$\begin{array}{c} CH_2-OH \\ | \\ C=O \\ | \\ CH_2-OPO_3H_2 \end{array}$$

Dihydroxyacetone phosphate

Figure 4.8 Fate of glycerol in the liver.

region of lower proton concentration. Thus, for every NADH + H^+, 3 ATP are formed. However, in the case of $FADH_2$, the first phosphorylation site is bypassed so that only 2 ATP are formed. Since for every glucose molecule 6 NADH + H^+ and 2 $FADH_2$ are formed, in the TCA cycle, there is a gain of 22 ATP.

4.2.4 OXIDATION OF FREE FATTY ACIDS

Only a small, albeit variable, amount of fat appears in the blood as FFA, with most being in the form of lipoprotein vesicles. The FFA are derived from stored triacylglycerol by lipolysis (figure 4.4). They are delivered in the blood to muscle, liver, heart, kidney and lung cells where they are broken down by β-oxidation to generate acetyl coA and NADH + H^+ and $FADH_2$, which enter the TCA cycle and the electron transport system respectively.

Following the hydrolysis of triacylglycerol the glycerol is transported to the liver, kidneys and intestines which have the enzyme glycerokinase. There, glycerol is phosphorylated at the expense of ATP to become glycerol-3-phosphate which is oxidized to form the triose phosphate dihydroxyacetone phosphate. Thus glycerol is able to enter the glycolytic pathway (figure 4.8).

The progressive removal of two carbon atoms from the fatty acid molecule is called β-oxidation (figure 4.9). The first reaction takes

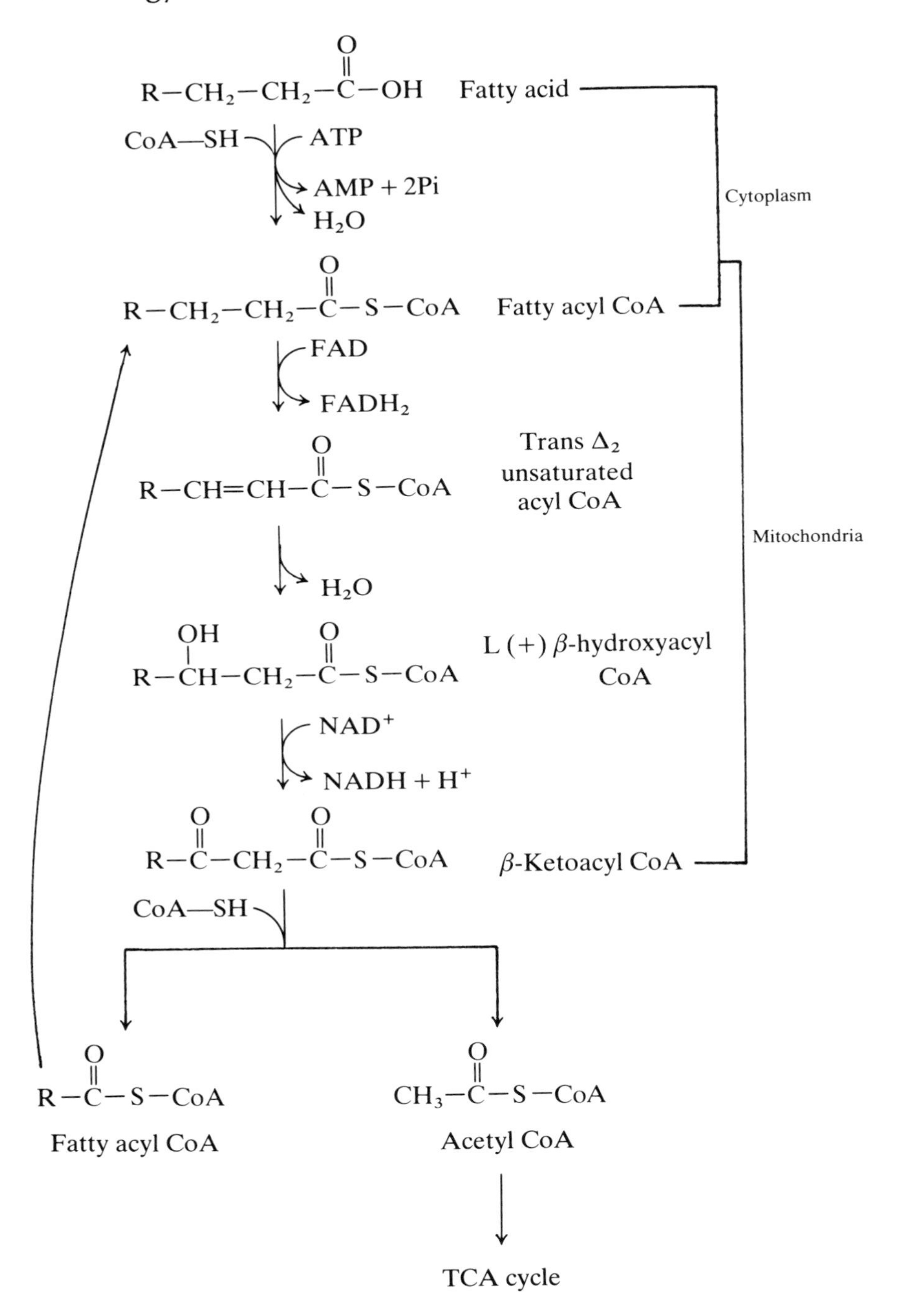

Figure 4.9 β-Oxidation of free fatty acids.

Table 4.3 Energy yield from β-oxidation and the TCA cycle for an 18C fatty acid (stearic acid)

β-Oxidation	
The fatty acyl coA cycle runs 8 times with 2C being removed each time, thus generating 9 acetyl coA	
8 $FADH_2$	16 ATP
8 $NADH + H^+$	24 ATP
TCA cycle	
The TCA cycle runs 9 times since 9 acetyl coA are produced	
9 × 1 $FADH_2$	18 ATP
9 × 3 $NADH + H^+$	81 ATP
9 substrate level	9 ATP

place in the cytoplasm of the cell with subsequent reactions occurring in the mitochondria. In this series of reactions, two pairs of electrons (1 × $FADH_2$ and 1 × $NADH + H^+$) are generated which subsequently enter the electron transport chain. The end-products of β-oxidation are fatty acyl coA and acetyl coA. Fatty acyl coA re-enters the pathway of β-oxidation until the fatty acid molecule is completely broken down whereas acetyl coA enters the TCA cycle. The amount of ATP generated from an 18-carbon fatty acid is summarized in table 4.3.

Therefore, a total of 148 ATP are generated. Since two phosphate groups from ATP are used in the first stage of β-oxidation, the net yield is 146 ATP.

Thus, per molecule of 18-carbon fatty acid, about four times as much ATP is generated compared with carbohydrate. When the ATP yield of a six-carbon fatty acid is compared with the ATP yield of glucose (also a six-carbon molecule), there is still a 20% higher ATP production from fat.

Fat also requires more oxygen per carbon atom for its breakdown than carbohydrate, and yields 10% less energy per molecule of oxygen consumed than does carbohydrate (table 4.4).

4.3 Anaerobic provision of ATP for muscular work

At the onset of exercise and during short-term exhaustive work, skeletal muscle fibres meet their ATP requirements initially (first 3–5 s) from the limited ATP and creatine phosphate pool within

Table 4.4 Oxygen consumption and ATP production

Glucose	ATP	O atoms
Glycolysis		
2 substrate level	2	–
$2 \times NADH + H^+$	6	2
Pyruvate → acetyl coA		
$2 \times NADH + H^+$	6	2
TCA cycle		
2 substrate level	2	–
$3 \times 2NADH + H^+$	18	6
$2 \times 1FADH_2$	4	2
Total	38 ATP	12 O (6 O_2)
Glucose generates 6.3 ATP per oxygen molecule consumed		
18-C fatty acid (stearic acid)		
β-Oxidation		
−2 ATP	−2	–
$8NADH + H^+$	24	8
$8FADH_2$	16	8
TCA cycle		
9 substrate level	9	–
$9 \times 3NADH + H^+$	81	27
$9 \times FADH_2$	18	9
Total	146 ATP	52 O (26 O_2)
An 18-C fatty acid generates 5.6 ATP per oxygen molecule consumed		

the fibres and subsequently from anaerobic glycolysis with the concomitant production of lactate. In very vigorous exercise, anaerobic metabolism ensures a rapid turnover of ATP which may exceed 1 mol min^{-1}. Muscle lactate production increases linearly with work intensity but the blood lactate concentration increases in a curvilinear fashion, with the onset of blood lactate accumulation occurring at a lactate concentration of about 2–4 mmol $litre^{-1}$. This point represents a rate of lactate production that exceeds the rate of lactate removal (figure 4.10). A blood lactate concentration in excess of 4 mmol $litre^{-1}$ is indicative of a significant increase in sympathetic nervous activity with a significant positive correlation between free plasma noradrenaline and muscle lactate concentration.

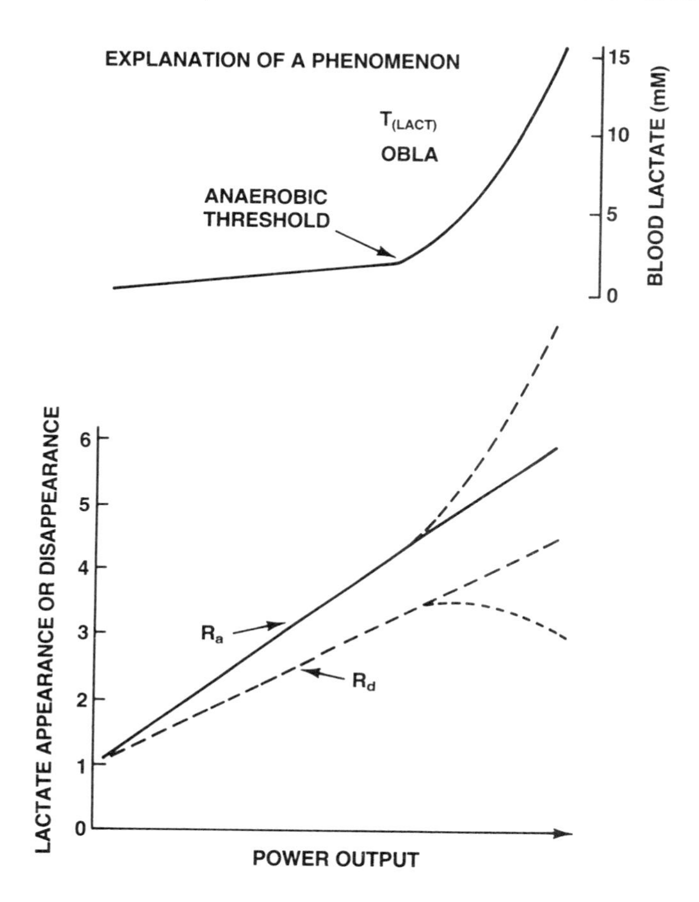

Figure 4.10 Onset of blood lactate accumulation. The accumulation of lactate in venous blood is a function both of the rate of lactate appearance (Ra) and the rate of lactate disappearance (Rd). As power output increases both Ra and Rd increase but the increase in Rd is less than that of Ra. Since neither the Ra or Rd is a linear function of power output, the blood lactate response to exercise is a curve. The point at which the blood lactate ceases to be a linear function of power output has been termed the "Anaerobic Threshold", although the terms "Onset of Blood Lactate Accumulation" (OBLA) or "Lactate Threshold" [T(lact)] are preferable. The term "Anaerobic Threshold" implies an oversimplistic explanation of the phenomenon. Recent studies suggest that as well as an increased rate of glycogenolysis at high work rates, lactate accumulates both due to a decreased Rd associated with a decrease in liver and kidney blood flow, and to an increased recruitment of IIb muscle fibres with their relatively poor capacity for aerobic metabolism. Figure reproduced with permission from G.A. Brooks (1985), © American College of Sports Medicine.

MUSCLE FATIGUE

During exercise in which anaerobic metabolism is the major source of ATP, there is a decrease in muscle pH. This decrease in pH interferes with the contractile process and energy metabolism, which subsequently limits the capacity of the muscle fibres to perform work. Under such circumstances muscle fatigue seems to be due to biochemical changes rather than a neural mechanism. However, during the first 60 s of maximum work when the motor units are all initially maximally active, there are electromyographic (EMG) changes which are indicative of neural fatigue.

4.4 Aerobic provision of ATP for muscular work

When the blood lactate concentration is less than about 4 mmol litre^{-1}, ATP is predominantly supplied by aerobic metabolism, and performance time is lengthened while the rate of work (power output) is reduced.

4.4.1 RELATIONSHIP BETWEEN OXYGEN CONSUMPTION AND AEROBIC METABOLISM

The percentage contribution of fatty acids and glucose can be estimated from measurements of the respiratory exchange ratio ($\dot{V}CO_2/\dot{V}O_2$) during submaximal exercise since the ratio of carbon dioxide produced to oxygen consumed reflects metabolism. The breakdown of glucose alone would result in a respiratory exchange ratio of 1.0 because carbon dioxide is produced in the same molecular amounts as oxygen is consumed:

$$-\underset{\underset{OH}{|}}{\overset{\overset{H}{|}}{C}}- + O_2 \longrightarrow CO_2 + H_2O = 1/1 = 1.0$$

In contrast the breakdown of fatty acids alone would result in a respiratory exchange ratio of 0.7:

$$-\underset{\underset{H}{|}}{\overset{\overset{H}{|}}{C}}- + 1.5\,O_2 \longrightarrow CO_2 + H_2O = 1/1.5 = 0.7$$

A mole of oxygen occupies 22.4 litres at standard temperature and pressure dry. Thus, from measurements of the number of litres of oxygen consumed, one can easily derive the number of moles of oxygen consumed (nO_2). For example an oxygen consumption of 1.0 litres per minute is equivalent to 44.6 mmol min^{-1} (1.0/22.4 mol min^{-1}). It will be shown in Chapter 5 that whole body oxygen consumption is not a measure of muscle oxygen consumption but that during vigorous exercise the muscles require the largest proportion of the total oxygen consumed. In the following example it is assumed that the active muscles require 75% of the total oxygen consumption.

An athlete has an $\dot{n}O_2$ max of 178.4 mmol min^{-1} (4.0 litres min^{-1}) and (if the muscles require 75% of this oxygen), a muscle $\dot{n}O_2$ max of 133.8 mmol min^{-1}. The athlete runs for 1 h at 50% of her $\dot{n}O_2$ max, so that her muscles are consuming 66.9 mmol of oxygen per minute. During the run she has a mean respiratory exchange ratio of 0.90. Using standard values of energy equivalents for a non-protein respiratory quotient (assuming RER = RQ), an RER of 0.90 indicates that she has derived two-thirds of her energy from carbohydrate and one-third from lipid (table 4.4). The total amount of energy (ATP) for a given oxygen consumption depends on the fuel mix, as indicated in tables 4.3 and 4.4, with more ATP being generated per mole of oxygen consumed for carbohydrate than for lipid. For an RER of 0.90, 6.13 mol of ATP are derived per mole of oxygen consumed (table 4.5).

Thus, in this example of a total skeletal muscle oxygen consumption of 4.014 mol of oxygen (66.9 mmol min^{-1} × 60 min) 24.6 mol of ATP are generated (4.014 mol O_2 × 6.13 mol ATP) of

Table 4.5 Energy equivalents for oxygen at different non-protein respiratory quotients

RQ	kJ litre^{-1}	Fuel contribution Lipid %	CHO %	Moles ATP produced per mole O_2 consumed
0.70	19.25	100.0	0.0	5.60
0.75	19.58	83.3	16.7	5.72
0.80	19.87	66.7	33.3	5.83
0.85	20.20	50.0	50.0	5.95
0.90	20.50	33.3	66.7	6.13
0.95	20.84	16.7	83.3	6.22
1.00	21.12	0.0	100.0	6.30

which 16.4 mol are derived from carbohydrate (66.7%) and 8.2 mol are derived from lipid (33.3%). If it is assumed that the carbohydrate is glucose (rather than glycogen), then 38 mol ATP are generated for every mole of glucose completely broken down and therefore 0.43 mol glucose must have been used to provide 16.4 mol ATP (16.4 mol ATP/38 mol glucose). Since glucose has a molecular weight of 180, this amounts to 77.4 g of glucose for the 1-h run (0.43 mol glucose × 180). If it is assumed that all the lipid used is in the form of 18-carbon fatty acids, then, since 146 mol ATP are generated per mole of 18-carbon fatty acid, 0.056 mol lipid are used, equivalent to 15.9 g as the molecular weight of an 18-carbon fatty acid is 284. These data are summarized in table 4.6, from which it can be seen that, although lipid contributes a third of the energy, it contributes only a sixth of the total weight loss of fuel and only a ninth of the total number of fuel molecules used.

The above calculation has been simplified since it does not take into account the protein oxidation rate. To do this it would be necessary to measure the urinary nitrogen excretion rate together with blood urea nitrogen in order to correct for any changes in the size of the body's amino acid pool during the period of measurement. For more detailed accounts of the calculation of fuel utilization from measurements of respiratory gas exchange and the limitations of the method see reviews by Frayn (1983), Ferrannini (1988), Elia and Livesey (1988) and Livesey and Elia (1988).

Two consequences of all this should now be apparent: first, that, for the continued provision of ATP, oxygen is needed since it is the final electron acceptor in the electron transport chain; and, secondly, a great deal of heat is being generated since ATP conserves only about three-quarters of the energy released in fuel breakdown. Thus,

Table 4.6 Summary of contributions of carbohydrate and lipid to fuel metabolism during exercise in a hypothetical 1-h run

ATP production (moles)		Weight of fuel (g)		Moles of fuel	
CHO	Lipid	CHO	Lipid	CHO	Lipid
16.4	8.2	77.4	15.9	0.43	0.056
67%	33%	82.9%	17.0%	88.6%	11.4%

the mechanisms involved in oxygen delivery to skeletal muscle (cardiovascular system) and heat dissipation (thermoregulatory mechanisms) are vital in exercise.

However, the immediate problem is to discuss how the autonomic nervous system contributes towards the control of cell metabolism.

4.5 The autonomic nervous system and metabolism

Parasympathetic neurones of the vagus nerve mediate hepatic glycogen synthesis while the sympathetic cholinergic fibres may have indirect effects on metabolism, for example on muscle metabolism during the first few seconds of activity, by moderating an increase in tissue blood flow. Noradrenergic sympathetic nerves have effects on tissue catabolism both directly and indirectly by influencing pancreatic hormone secretion during exercise. Circulating adrenaline plays an important role in stimulating muscle and liver glycogenolysis while the noradrenergic neurones are important in stimulating adipose tissue lipolysis.

4.5.1 MUSCLE GLYCOGEN

During short-duration exercise muscle glycogen breakdown increases due to an increase in local Ca^{2+} concentration that occurs as a result of motor nerve stimulation. This causes the conversion of inactive glycogen phosphorylase *b* to the active form, glycogen phosphorylase *a* (figure 2.5). However, there is a dual role for motor nerve activity and adrenaline, with adrenaline having an important role in prolonged exercise. Using an isolated rat hind limb preparation, Richter *et al.* (1982) showed that adrenaline stimulates muscle glycogenolysis during intense electrical stimulation over 20 min in slow twitch fibres but not in fast twitch fibres. However, during low-intensity muscle stimulation, adrenaline stimulates muscle glycogenolysis in fast twitch fibres but not in slow twitch fibres. The authors suggest that the findings may be due to the ability of glycogenolysis to proceed in fast twitch fibres during rapid contractions, with adrenaline being needed to enhance FFA supply to the slow twitch muscle. In contrast, during low-intensity stimulation, adrenaline stimulates glycogenolysis in fast twitch muscle but has no effect on slow twitch muscle since in the latter fibres the metabolic demands of low-intensity stimulation are met without the need for stimulation of metabolism by adrenaline.

Other studies have used adrenaline infusions to demonstrate a role for adrenaline in muscle glycogen breakdown. However, caution should be applied when the concentrations of adrenaline in blood exceed the normal physiological range. For example, in one study adrenaline was infused into the femoral artery of one leg during 45 min of submaximal exercise. This caused an increase in muscle glycogenolysis in the treated leg. The authors estimated that the blood levels reached in this leg were 12.7 nmol $litre^{-1}$, and suggested that they are physiological (Jansson *et al.*, 1986). However, compared with the maximal values reported in a variety of physiological and pathophysiological states, this value is pathophysiological.

4.5.2 LIVER GLYCOGEN

In man liver glycogenolysis can be stimulated via the β_2-adrenoceptor-mediated adenylate cyclase system or by an α-adrenoceptor-mediated mechanism involving an increase in local Ca^{2+} levels. However, many studies of exercise metabolism have used rats, a species which has no functional β-adrenoceptors in liver and therefore any sympathoadrenal effects are mediated by α-adrenoceptors. The liver is the only source of blood glucose, which is formed from glycogenolysis and from gluconeogenesis. Muscle glycogen cannot contribute to blood glucose since muscle lacks the enzyme glucose-6-phosphatase (figure 4.5). The main pathway that produces glucose for the blood is glycogenolysis. However, some glucose is derived from gluconeogenesis when the three-carbon precursors such as lactate and glycerol are available. The catecholamines are involved in the provision of lactate and glycerol to the liver since they mediate muscle glycogenolysis and adipose tissue lipolysis. Gluconeogenesis is stimulated by an adrenoceptor-mediated mechanism which does not require the cyclic AMP cascade system.

Adrenodemedullation has no effect on liver glycogenolysis in rats after 5 or 30 min of high intensity exercise (Marker *et al.*, 1986) or after 30 or 60 min submaximal exercise (Carlson *et al.*, 1985). Thus, the latter authors suggest that glucagon stimulation of hepatic cyclic AMP or autonomic nervous stimulation mediates liver glycogenolysis during exercise. Carlson *et al.* (1985) claim that the discrepancy between their results and the decrease in liver glycogen content seen with previous studies using adrenodemedullation can be explained by the animal handling procedures. The authors suggest that irregular handling of the animals and exposing them to exercise for the first time during the experiment itself led to stress in rats of other experiments. The stress thus imposed on the animals could have raised the adrenaline levels sufficiently to stimulate hepatic glycogenolysis. However, Sonne *et al.* (1985), also using rats accustomed to exercise, came to the opposite conclusion. They surgically denervated the liver or adrenodemedullated rats to compare the effects of circulating catecholamines versus the noradrenergic neurones. Unlike Carlson *et al.* (1985) they also measured glucose production and removal rates. In both the control (sham operated) and denervated rats, blood glucose increased during moderate exercise since hepatic glucose production exceeded the tissue use of glucose (figure 4.11). There was no difference in this regard between the two groups of rats. Neither was there any

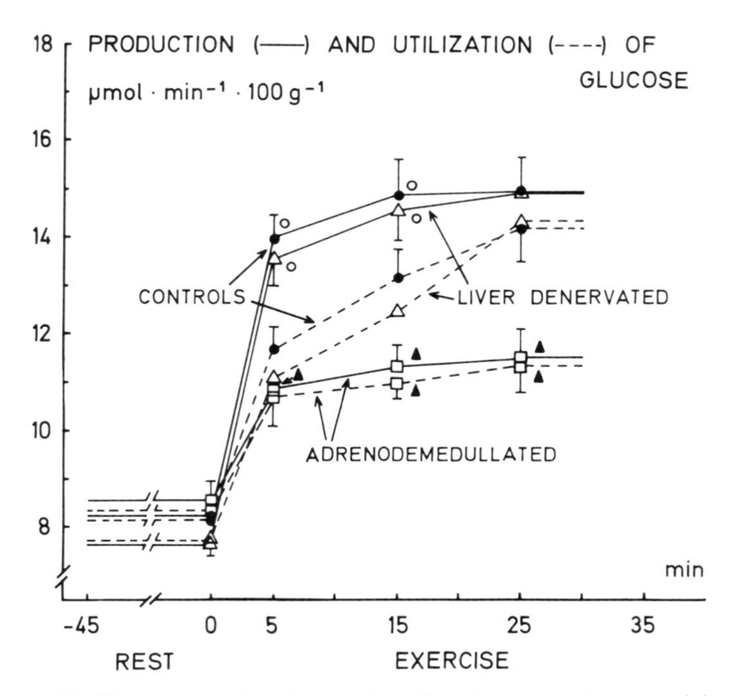

Figure 4.11 Glucose production and utilization rates in exercising rats. Figure reproduced with permission from Sonne *et al.* (1985).

difference in hepatic glycogen depletion. Thus, hepatic innervation has no role in glucose turnover either at rest or during moderate exercise in rats (although a small effect of hepatic sympathetic nervous activity on glucose production is seen in man at rest). However, in the adrenodemedullated rats, hepatic glucose production was attenuated by 50% of the control value and was the same as glucose uptake in this group of rats. Thus, there was no change in blood glucose or hepatic glycogen content. One reason for the differences between the two studies could be the exercise regimen. Carlson *et al.* exercised their rats for 60 min using a treadmill speed of 21 m min^{-1} with an incline of 15%. Sonne *et al.* worked their rats less hard, using the same speed (21 m min^{-1}) but with a level treadmill and for only 35 min.

4.5.3 LIPID

Lipolysis in adipose tissue is a β-adrenoceptor- (predominantly β_1) mediated mechanism involving the production of cyclic AMP and activation of an enzyme cascade. However, *in vitro* studies of human adipocytes taken from hypothyroid individuals show that there is also a mechanism for α-adrenoceptor-mediated inhibition of lipolysis, which occurs during fasting. During exercise the rate of lipolysis is related to circulating catecholamine levels although plasma catecholamines are not necessarily the stimulus: noradrenaline is simply an index of sympathetic nervous activity. Evidence for a causal relationship comes from studies using selective (β_1) and non-selective (β_1 and β_2) antagonism. Such studies show similar reductions in circulating FFA with a β_1-adrenoceptor antagonist as with a non-selective one. In man noradrenaline from sympathetic nerve terminals is much more important than adrenomedullary adrenaline since FFA and glycerol levels in blood increase during exercise after adrenalectomy in patients with Cushing's disease. Since plasma FFA and glycerol still increase (albeit significantly less) during adreno-ceptor antagonism, it seems that mechanisms of control, in addition to the sympathetic nervous system, are involved. Alternatively, this increase in plasma FFA and glycerol, despite β-adrenoceptor antagonism, could be explained by the fact that β-adrenoceptor antagonists are competitive, such that, if the stimulus for adreno-ceptor-mediated effects is increased, the antagonism will diminish. However, caution should be applied in interpreting measurements of FFA as an indicator of lipolysis. The FFA concentration in blood is affected by the rate of release of FFA into blood, the rate of FFA uptake by the cells and tissue blood flow. To some extent the measurement of plasma glycerol is a preferable index of lipolysis although this too is affected by tissue perfusion and has a very different volume of distribution from FFA. In addition, the altered use of glycerol for gluconeogenesis will affect the plasma glycerol concentration.

4.5.4 INSULIN AND GLUCAGON

The other principal controlling mechanisms of cell metabolism during exercise are the pancreatic hormones. These are in turn influenced by the catecholamines. Insulin is an anabolic hormone causing an increase in glycogen synthesis and lipogenesis, effects which are inappropriate during exercise. Catecholamines suppress insulin secretion via an α_2-adrenoceptor mechanism, although in some circumstances the catecholamines will increase insulin secretion via a β-adrenoceptor mechanism. Glucagon release is enhanced via a β-adrenoceptor mechanism. Glucagon, like adrenaline, causes an increase in liver glycogenolysis and gluconeogenesis via the cyclic AMP cascade system. It has an important role in preventing the vasoconstrictor effects of the catecholamines in the hepatic vasculature. For a review of the hormonal control of metabolism during exercise see Galbo (1983).

References and further reading

Barwich, D., Hägele, H., Weiss, M. and Weicker, H. (1981) Hormonal and metabolic adjustment in patients with central Cushing's disease after adrenalectomy. *Int. J. Sports Med.*, **2**, 220–227.

Brooks, G.A. (1985) Anaerobic threshold: Review of the concept and directions for future research. *Med. Sci. Sports Exer.*, **17**, 22–31.

Carlson, K.I., Marker, J.C., Arnall, D.A., Terry, M.L., Yang, H.T., Lindsay, L.G., Bracken, M.E. and Winder, W.W. (1985) Epinephrine is unessential for stimulation of liver glycogenolysis during exercise. *J. Appl. Physiol.*, **58**, 544–548.

Cryer, P.E. (1980) Physiology and pathophysiology of the human sympathoadrenal neuroendocrine system. *New Engl. J. Med.*, **303**, 436–444.

Elia, M. and Livesey, G. (1988) Theory and validity of indirect calorimetry during net lipid synthesis. *Am. J. Clin. Nut.*, **47**, 591–607.

Felig, P. and Koivisto, V. (1979) The metabolic response to exercise: implications for diabetes. *Therapeutics through Exercise* (eds Lowenthal, D.T., Bharadwaja, K. and Oaks, W.W.), W.B. Saunders, Orlando, Fla, pp. 3–20.

Ferrannini, E. (1988) The theoretical bases of indirect calorimetry: a review. *Metabolism,* **37**, 287–301.

Frayn, K.N. (1983) Calculation of substrate oxidation in vitro from gaseous exchange. *J. Appl. Physiol.,* **55**, 628–634.

Galbo, H. (1983) *Hormonal and Metabolic Adaptation to Exercise*, Georg Thieme Verlag, Stuttgart.

Hutson, N.J., Brumley, E.T., Assimacopoulos, F.D., Harper, S.C. and Exton-Smith, J.H. (1976) Studies of the α adrenergic activation of hepatic glucose output. I: Studies on the α adrenergic activation of phosphorylase and gluconeogenesis and inactivation of glycogen synthase in isolated rat liver parenchymal cells. *J. Biol. Chem.* **251**, 5200–5208.

Jansson, E., Hjemdahl, P. and Kaisjer, L. (1986) Epinephrine-induced changes in muscle carbohydrate metabolism during exercise in male subjects. *J. Appl. Physiol.,* **60**, 1466–1470.

Karlsson, J. (1985) Metabolic adaptations to exercise: a review of potential beta-adrenoceptor antagonistic effects. *Am. J. Cardiol.,* **55**, 48D–48D.

Koch, G., Franz, I.-W., Gubba, A. and Lohman, F.W. (1983) β Adrenoceptor blockade and physical activity: cardiovascular and metabolic aspects. *Acta Med. Scand.* Suppl., **672**, 55–62.

Livesey, G. and Elia, M. (1988) Estimation of energy expenditure, net carbohydrate utilization, and net fat oxidation and synthesis by indirect calorimetry: evaluation of errors with special reference to the detailed composition of foods. *Am. J. Clin. Nutr.,* **47**, 608–628.

Macdonald, I.A., Bennett, T. and Fellows, I.W. (1985) Catecholamines and the control of metabolism. *Clin. Sci.,* **68**, 613–619.

Marker, J.C., Arnall, D.A., Conlee, R.K. and Winder, W.W. (1986) Effect of adrenomedullation on metabolic responses to high-intensity exercise. *Am. J. Physiol.,* **251**, R552–R559.

Richardson, P.D.I. and Withrington, P.G. (1976) The inhibition by glucagon of the vasoconstrictor actions of noradrenaline, angiotensin and vasopressin on the hepatic arterial vascular bed of the dog. *Br. J. Pharmacol.,* **57**, 93–102.

Richardson, P.D.I. and Withrington, P.G. (1977) Glucagon inhibition of hepatic arterial responses to hepatic nerve stimulation. *Am. J. Physiol.,* **233**, H647–H654.

Richter, E.A., Ruderman, N.B., Gavros, H., Belur, E.R. and Galbo, H. (1982) Muscle glycogenolysis during exercise: dual control by epinephrine and contractions. *Am. J. Physiol.*, **242**, E25–E32.

Sonne, B., Mikines, K.J., Richter, E.A., Christensen, N.J. and Galbo, H. (1985) Role of liver nerves and adrenal medulla in glucose turnover of running rats. *J. Appl. Physiol.*, **59**, 1640–1646.

Stainsby, W.N., Sumners, C. and Eitzman, P.D. (1985) Effects of catecholamines on lactic acid output during progressive working contractions. *J. Appl. Physiol.*, **59**, 1809–1814.

Tate, C.A., Scherer, N.M. and Stewart, G. (1986) Adrenergic influence on hormonal and hepatic metabolic responses to exercise in rats. *Am. J. Physiol.*, **250**, R1060–R1064.

Tolbert, M.E.M., Butcher, F.R. and Fain, J.N. (1973) Lack of correlation between catecholamine effects or cyclic adenosine 3′ : 5′-monophosphate and gluconeogenesis in isolated rat liver cells. *J. Biol. Chem.*, **248**, 5686–5692.

Chapter 5

The cardiovascular and respiratory systems

In Chapter 4 it was shown that muscle requires oxygen for the continued generation of ATP from metabolism to enable muscle contraction. This oxygen is delivered from the lungs to the skeletal muscles via the cardiovascular system (i.e. the heart and blood vessels). This chapter is divided into two sections, A, dealing with cardiovascular responses to exercise, and B, examining respiration during exercise. The involvement of the autonomic nervous system in their control is also discussed.

A: CARDIOVASCULAR RESPONSES TO EXERCISE

5.1 The heart

5.1.1 STRUCTURE AND FUNCTION

The right and left sides of the heart act as two separate systems for pumping blood, with the right side pumping blood through the pulmonary circulation and the left side pumping blood through the systemic circulation (figure 5.1). Systemic venous blood which is low in oxygen as it drains the tissues enters the right atrium through the superior and inferior venae cavae. Most of this blood flows through the tricuspid valve directly into the right ventricle. However, an additional 30% of blood enters the right ventricle as a result of subsequent atrial contraction. Blood fills the right ventricle, and an increase in ventricular pressure causes the tricuspid valve to close. When the tricuspid valve closes, the right atrium fills with blood from

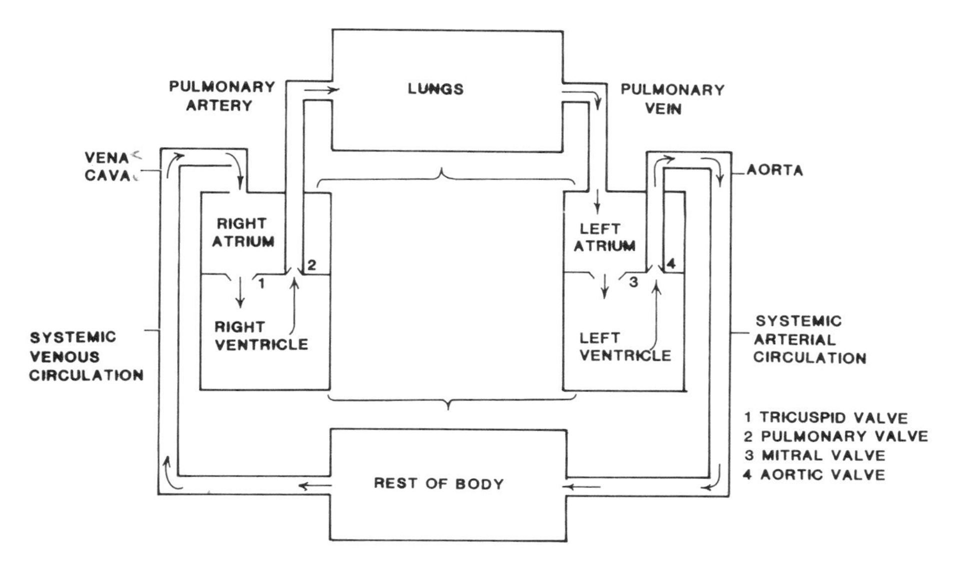

Figure 5.1 The pulmonary and systemic circulations.

the vena cava as this blood is no longer able to pass freely into the right ventricle, and thus right atrial pressure increases. Ventricular contraction brings about an increase in right ventricular pressure, which ultimately causes opening of the pulmonary valve that separates the right ventricle from the pulmonary artery, and blood is ejected under pressure from the right ventricle through the pulmonary circulation. As soon as the period of ventricular contraction (**systole**) is over, the pressure falls in the right ventricle. The increased pressure in the right atrium, described above, causes the tricuspid valve to open and the **cardiac cycle** (cycle of atrial and ventricular contraction) is repeated.

The blood which leaves the right ventricle enters the pulmonary circulation, passes through capillaries around the alveoli and returns to the left atrium via the pulmonary veins rich in oxygen. The cardiac cycle in the left side of the heart is the same as that occurring in the right-hand side of the heart. Thus, blood flows through the left atrium freely to the left ventricle via the mitral valve. Again a further 30% of blood enters the ventricle due to left atrial contraction. Blood leaves the left ventricle through the aortic valve into the systemic circulation. The left ventricle pumps the same volume of blood per minute as the right ventricle but at about six times the pressure. This greater pressure is necessary since the systemic circulation offers much more resistance than the pulmonary circulation. The development of the greater pressure which is required in the left ventricle is facilitated by a thicker muscular wall in the latter ventricle.

Each cardiac cycle lasts for 0.8 s in a resting individual, with ventricular systole lasting 0.3 s and ventricular filling time or period of ventricular relaxation (**diastole**) lasting for 0.5 s. This series of events occurs simultaneously in the two sides of the heart.

5.1.2 HEART SOUNDS

The simultaneous closure of the tricuspid and mitral valves, together with blood turbulence, cause the first heart sound, which therefore marks the onset of ventricular systole. The subsequent simultaneous closure of the pulmonary and aortic valves is the main cause of the second heart sound and marks the onset of ventricular diastole. There is a longer delay between the second and first heart sounds than between the first and second due to the greater amount of time taken for diastole compared with systole. The first heart sound is conventionally referred to as 'lub' and the second, shorter and harder, sound is known as 'dub'.

5.1.3 CARDIAC MUSCLE

The atrial and ventricular walls are composed of cardiac muscle cells, which are rich in mitochondria necessary to provide ATP for the continually active cells. There is a dense capillary network, with around one capillary per muscle fibre which facilitates the diffusion of oxygen, carbon dioxide and waste products. The cardiac muscle cells appear striated because, like skeletal muscle cells, they are composed of sarcomeres of actin and myosin filaments. However, the cardiac muscle cells, unlike skeletal muscle cells, are not continuous along the full length of the heart. The cells are relatively short (100 μm long) and joined end to end by intercalated discs. The discs offer little resistance to an action potential which is therefore readily transmitted from one cardiac muscle cell to the next. Thus, cardiac muscle acts as a **syncytium** (mass of protoplasm containing many nuclei). This means that cardiac muscle cells, like unitary smooth muscle cells (see section 5.2.3), contract as a unit. They do so spontaneously, and this continues even when they are removed from the body. Thus, no central command is required from the rhythmic contraction of heart muscle. The role of the autonomic nervous system is to modulate this inherent rhythmicity.

A distinct electrical feature of the cardiac cell is the relatively slow rate at which it is depolarized and the relatively long length of time of the depolarization. This is due to the entry of Ca^{2+} as well as Na^{+} ions into the cell, mainly via slow channels, which does not occur in other excitable tissues.

There are two fundamental ways in which the strength of cardiac muscle contraction is controlled. The first is the intrinsic **Frank–Starling mechanism**, and the second involves autonomic nerves affecting **inotropic** (force) and **chronotropic** (rate) cardiac function. These mechanisms are discussed in section 5.4.2.

5.1.4 AUTONOMIC INNERVATION OF THE HEART

The heart beats in response to electrical stimulation originating in the **sinoatrial (SA) node** of the right atrium. The SA node receives parasympathetic and sympathetic nerve fibres. The parasympathetic neurones originate in the so-called **cardiac centre** of the medulla. These medullary neurones make connections with the vagus nerve (cranial nerve X) which originates in the medulla (see Chapter 1). The vagus nerve passes through the neck to the heart where its branches synapse in ganglia near the SA and atrioventricular (AV) nodes. The right branch of the vagus nerve predominantly supplies the SA node and the left branch predominantly supplies the AV node. The vagus nerve slows down the rate of rise of the action potential of the pacemaker cell since acetylcholine causes hyperpolarization as a result of an increased permeability of the membrane to potassium.

The sympathetic nerve fibres have their origin in a T1–6 and C1–2 in the spinal cord (see Chapter 1). The postganglionic cells innervate the myocardial muscle cells as well as the SA and AV nodes. On the basis of tissue noradrenaline concentration, the atrial cells are about three times more densely innervated than the ventricular cells. The sympathetic noradrenergic fibres act on the SA node in the opposite manner to the vagal cholinergic fibres, increasing the rate of rise of action potential of the pacemaker cell. In addition to increasing the rate of cardiac contraction (chronotropic action), sympathetic neurones are able to increase the strength of cardiac contraction (inotropic action). These neurones control the force of contraction by a β_1-adrenoceptor-mediated mechanism involving cyclic AMP as the second messenger (see Chapter 2). This results in the faster movement of Ca^{2+} into the cell during depolarization. Once inside the cell Ca^{2+} ions bind to troponin and are thereby involved in excitation–contraction coupling as in skeletal muscle (see Chapter 3).

Parasympathetic and sympathetic neurones have an interactive effect on the heart which cannot be explained in a simple additive way. The effect that vagal stimulation has on the heart depends on the degree of existing sympathetic activity. If there is pre-existing sympathetic activity, the inhibitory effect of vagal activity will be augmented. This is known as **accentuated antagonism**. It seems to be due to acetylcholine released from the parasympathetic vagal neurones causing a decrease in the rate of noradrenaline release from adjacent sympathetic nerve terminals. This could theoretically be an important way of keeping the heart rate low – and therefore maximizing cardiac filling time – during exercise since only at maximal exercise is vagal tone completely lost. However, at present there is no direct evidence that such a mechanism operates during exercise.

5.1.5 ELECTRICAL ACTIVITY IN THE HEART

The SA node is electrically connected to the AV node by three bundles of internodal pathways. When electrical activity passes along these conducting fibres and spreads into the adjacent muscle cells, the atrial cells are depolarized and thereby they are stimulated to contract. The electrical impulse is briefly held up at the AV node. This impulse delay precedes the completion of atrial ejection of blood into the ventricles. The atrial cells repolarize as the electrical impulse continues through the heart along further specialized conducting fibres in the ventricles, the **bundle of His** and the **Purkinje fibres**. The spread of electrical activity into the ventricles precedes their mechanical contraction. The ventricular cells then repolarize in the reverse order from which they were depolarized. Once the ventricular mass is fully repolarized, a new wave of excitation is initiated at the SA node. Thus, although the electrical and mechanical events that comprise a cardiac cycle are discrete, they are temporally coupled with a time delay. The time delay is explained by the fact that electrical activity is necessary to initiate muscle contraction.

The electrocardiogram

The electrical activity in the heart leaks through the body tissues, which are composed of around 60% water (and are therefore good electrical conductors), to the skin. The electrical events can therefore be monitored by surface recording electrodes using an **electrocardiogram (ECG).** The signals are amplified and displayed either on an oscilloscope or chart recorder as an **electrocardiograph**. The electrical picture obtained depends on the location of the recording electrodes. There are 12 conventional configurations of electrodes, with each configuration involving three or four electrodes. For clinical purposes it is essential that the standard electrode positions be adhered to, and that the electrical signal be calibrated so that a 1 cm deflection on the electrocardiograph corresponds to 1 mV of electrical signal. However, if the ECG is used simply to monitor heart rate, or to detect arrhythmias and abnormal waveforms by physiologists without the specialist knowledge or requirements of the cardiologist, such precision is not necessary. Rather, it is important only to obtain a trace in which all the waveforms are clearly evident.

The fundamental principles of electrocardiography can be illustrated using three electrodes of which one is an earth, one an exploring or recording electrode and one an indifferent electrode which is therefore held at zero potential. Such a combination produces a unipolar recording since only one electrode is used to record the heart's electrical activity. The unipolar leads can be placed either on the wrists and ankles (**augmented limb leads**) or on the chest (**praecordial leads**). For routine monitoring of heart rate, rhythm and waveform during exercise it is easier to use praecordial leads. In this case, theoretically the earth electrode can be placed anywhere on the skin but it is usually convenient to place it on the back of a shoulder or at the base of the sternum. The indifferent electrode is placed above the heart, typically on the sternum where there is little muscle mass and therefore movement which could interfere with the electrical signal. The recording electrode is placed below the level of the heart. The size of the waveform depends on the position of this electrode, with larger waves occurring as the electrode is moved from the sternum to below the axilla on the left side at a level approximately corresponding to the fifth rib. This is because the electrode is being moved closer to the main vector of depolarization through the heart. The potential difference between the recording electrode and the indifferent electrode in such a combination produces a positive (upward) deflection on the ECG as the wave of cardiac excitation moves towards the recording electrode. Clearly, if the indifferent and recording electrodes were reversed, the wave of electrical activity would move away from the recording electrode and the main wave of cardiac excitation would produce a large negative deflection on the ECG.

The waveforms in the ECG are arbitrarily assigned the letters P, Q, R, S and T. The first wave, the P wave, is due to the wave of electrical activity through the atria from the SA to AV node. With praecordial electrodes in the positions described above, this is seen as a positive deflection. As the electrical signal is held up at the AV node there is a period of electrical silence. This is the P–R interval. The mass of electrical activity that passes along the bundle of His and Purkinje system and leaks out into the surrounding cardiac muscle cells gives rise to the QRS complex of waves which displays a prominent upward deflection of the R wave when the recording electrode is over the base of the heart. The wave of repolarization of the atria is masked by the QRS complex. Between the S wave and the start of the next P wave the ventricles repolarize. This includes the T wave which is a positive deflection since the ventricular cells that depolarized last are the first to repolarize (figure 5.2).

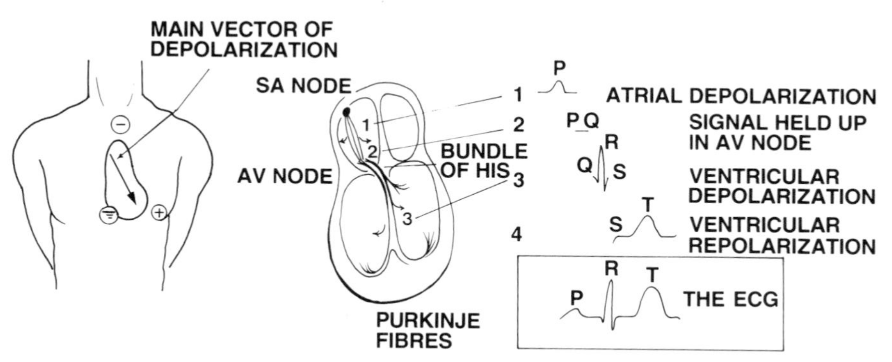

Figure 5.2 The electrocardiograph.

Since the electrical events in the heart result in the mechanical events of cardiac contraction (**systole**) and relaxation (**diastole**), the ECG can be used to determine their time course. Thus, it can be easily demonstrated that both the times of systole and diastole shorten as heart rate increases during exercise. However, it is important to recognize that it is diastole that is compromised the most (figure 5.3). The dramatic reduction in time for cardiac filling during exercise has wider implications which are discussed later.

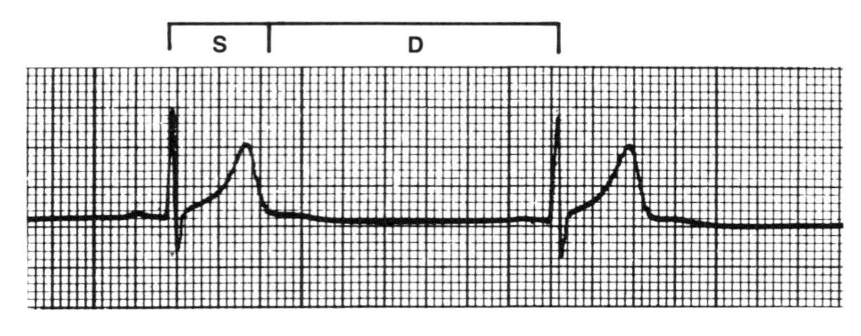

(a)

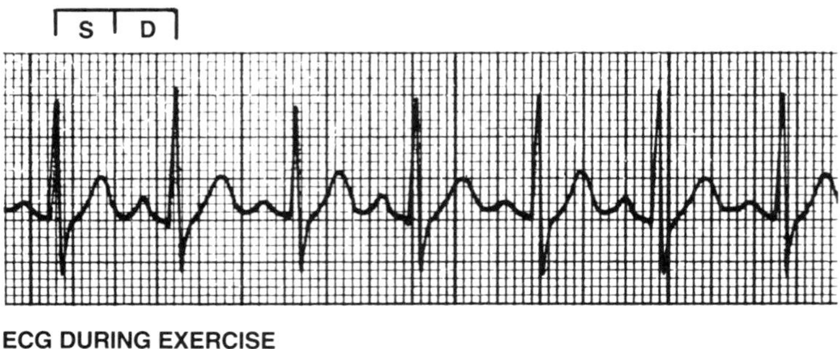

(b)

1 SEC

Figure 5.3 The ECG at rest and during exercise in a well-trained runner. (a) Resting ECG. Heart rate = 32 b min^{-1}; ventricular systole (S) = 0.48 s; ventricular diastole (D) = 1.40 s. (b) Exercising ECG. Heart Rate = 107 b min^{-1}; ventricular systole (S) = 0.28 s; ventricular diastole (D) = 0.28 s.

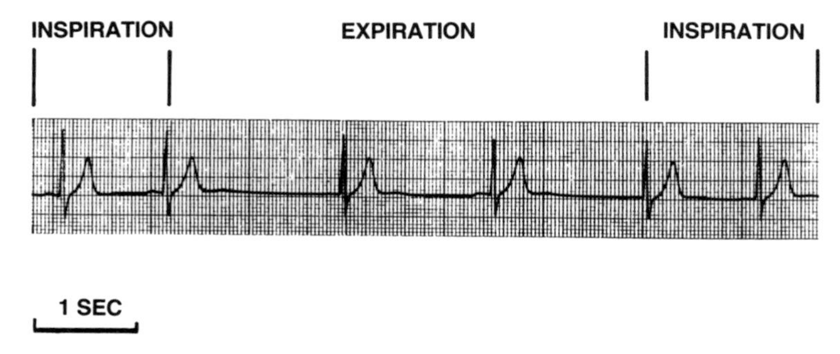

Figure 5.4 Sinus arrhythmia in a young adult athlete. Heart rate during inspiration = 53 b min^{-1}; heart rate during expiration = 38 b min^{-1}.

Sinus arrhythmia

In resting healthy young adults the heart rate varies with respiration. When the inspiratory neurones fire, vagal motor activity to the heart is decreased leading to an increase in heart rate (figure 5.4). Several mechanisms may explain this phenomenon. To some extent it may be due to a direct influence of the medullary inspiratory neurones on the vagal neurones supplying the SA node. Other mechanisms include the **stretch reflex** in the lungs and the **Bainbridge reflex**.

Stretch reflex During inspiration the stretch receptors in the lungs reflexly inhibit vagal discharge from the medulla so that heart rate increases. During expiration, when the stretch receptors are no longer stimulated, the heart rate decreases due to vagal inhibition of the SA node. Thus, the degree of sinus arrhythmia in an individual at rest is an indicator of vagal integrity. During exercise the sinus arrhythmia disappears as vagal tone is reduced.

Bainbridge reflex An increase in right atrial pressure causes firing of the atrial stretch receptors. This information is relayed via the vagus nerve to the medulla. Efferent signals are relayed back to the heart, again via the vagus nerve, and reflexly cause an increase in the rate and force of cardiac contraction known as the Bainbridge reflex. During inspiration the decrease in intrathoracic pressure causes a change in venous pressure sufficient to increase venous return to the atria and initiate this stretch reflex.

5.2 Blood flow

5.2.1 POISEUILLE'S LAW

Before discussing blood flow the relevant basic properties of fluid dynamics will be outlined. Fluid flow through tubes may be either **laminar** (that is, streamlined) or **turbulent**. The rate of laminar fluid flow through a tube can be determined mathematically using **Poiseuille's law of laminar flow**. Since blood flow through the vasculature is nearly always laminar, the law is pertinent to an understanding of blood flow.

Four factors affect laminar flow:

1. Flow is proportional to the fluid pressure difference along the tube:

$$Q \propto P_1 - P_2,$$

where Q = flow, P_1 = pressure of fluid entering the tube, P_2 = pressure of fluid leaving the tube.

2. Flow is proportional to the tube radius raised to the fourth power:

$$Q \propto r^4.$$

3. Flow is inversely proportional to the length of the tube:

$$Q \propto 1/l.$$

4. Flow is also affected by the viscosity of the fluid (n).

These four factors are expressed in Poiseuille's law:

$$Q = \frac{\pi(P_1 - P_2)\, r^4}{8nl}$$

where $\pi/8$ is constant.

5.2.2 FACTORS DETERMINING BLOOD FLOW

Poiseuille's law can be simplified when it is applied to the cardiovascular system because blood vessel lengths are constant and blood viscosity is virtually constant. The latter decreases by about 2.6% per degree Celsius increase in blood temperature, and increases with an increase in haematocrit, typical after acclimation to high altitude. However, changes in blood flow are principally brought about through changes in arterial blood pressure and vascular resistance (R). Thus,

$$Q = \frac{P_1 - P_2}{R}$$

$$R = \frac{P_1 - P_2}{Q}$$

$$P_1 - P_2 = Q \times R$$

In the context of cardiac output, P_1, is mean arterial blood pressure, i.e. the driving pressure provided by the heart. (Mean arterial blood pressure is the diastolic blood pressure plus a third of the pulse pressure (difference between diastolic and systolic blood pressure).) Thus, a resting individual with a systolic blood pressure of 16 kPa (120 mm Hg) and a diastolic blood pressure of 10.6 kPa (80 mm Hg) has a mean blood pressure of 12.4 kPa (93 mm Hg) (i.e. $10.6 + (0.33 \times 5.4) = 12.4$). P_2 is the venous pressure as blood re-enters the heart, and this pressure is near to zero. Blood pressure decreases as blood travels through the systemic or pulmonary circulation from the arteries, arterioles, capillaries and venules and passes finally to the veins. These pressure differences are essential for flow to occur. However, the major determinant of blood flow through any particular tissue is vascular resistance, which is governed by the blood vessel radius in the microcirculation.

5.2.3 THE VASCULATURE

The arteries have thick walls composed of elastic tissue and some smooth muscle cells. The high elastic content allows arteries to transport blood at high pressure. The arteries branch into smaller arterioles which contain relatively more smooth muscle than the arteries. This allows the arterioles to constrict and reduce blood flow or dilate and increase blood flow. The arterioles are therefore also termed the **resistance vessels**. Blood from the arterioles passes into the thin-walled capillary vessels. It is in the capillary bed that blood gases, nutrients and waste products are exchanged between the blood and tissues. Blood returns to the heart via the venules and larger veins which act as capacitance vessels since the larger portion of the blood volume is contained in these vessels. The venous blood pressure is normally low but smooth muscle in the vein walls allows them to constrict to increase this venous pressure and augment venous return (or dilate and reduce pressure). Reflex venoconstriction in the veins draining active muscle during exercise aids venous return to the heart.

Vascular resistance

Arteriolar vascular resistance is modified by neural and local factors. The extent of involvement of each of these factors depends on the vascular bed in question. In the skin and splanchnic beds the autonomic nervous system plays the major role, whereas in cardiac and skeletal muscle local factors are pre-eminent.

The main portion of the arterioles is smooth muscle which is innervated by sympathetic postganglionic nerve fibres. Smooth muscle cells are not striated like skeletal muscle because their protein filaments are distributed irregularly in parallel bundles. There are two main types of smooth muscle, **unitary** and **multiunit.**

Unitary smooth muscle cells respond to neural stimuli as a unit. They are found in the viscera, uterus, ureter, small arteries and veins. The cells are often spontaneously active with the autonomic nervous system serving to modulate their activity rather than initiate it. Multiunit smooth muscle cells are found in the large arteries and some regions of the intestinal tract. The activity of these cells is initiated by the sympathetic autonomic nervous system, facilitated by a rich supply of adrenoceptors in the cell membrane.

Autonomic innervation of the vasculature

Sympathetic vasoconstrictor fibres arise from the medulla and synapse in the spinal cord at the level of T1–L3 before reaching ganglia either in the sympathetic chain or in one of the other sympathetic ganglia (see Chapter 1). Noradrenaline released from the postganglionic nerve terminals causes vasoconstriction by acting on α-adrenoceptors in the vasculature of the arterioles (resistance vessels) and veins (capacitance vessels).

The arterioles of skeletal muscle and some areas of skin are also innervated by sympathetic cholinergic fibres which cause vaso-dilatation. Sympathetic cholinergic activity is thought to increase muscle blood flow during the initial 10 s or so of exercise before local factors have had time to take effect. However, such an early rise in blood flow, resulting in an immediate increase in oxygen delivery to the active muscle cells, would not be consistent with the well-established initial reliance on the anaerobic provision of ATP. The skeletal muscle arterioles also contain β_2-adrenoceptors which are capable of mediating vasodilatation although this mechanism is not thought to be of functional significance during exercise.

Neither muscle nor skin blood vessels receive parasympathetic neurones. However, cranial parasympathetic fibres, including the vagus nerve originating in the medulla, innervate cerebral and certain visceral blood vessels. Sacral parasympathetic fibres innervate blood vessels of the urogenital system. Stimulation of these fibres causes vasodilatation. However, the role of parasympathetic neurones in the control of arterial blood pressure is less important than the role of the sympathetic noradrenergic fibres.

Factors affecting the medullary neurones in the control of vascular resistance The medullary neurones, which have such a powerful influence over the neural control of vascular resistance, are themselves affected by afferent neural signals arising from atrial receptors, baroreceptors, chemoreceptors, cells of the hypothalamus and cerebral cortex, and cutaneous thermoreceptors. Temperature-sensitive neurones in the skin and hypothalamus mediate vasodilatation or vasoconstriction, depending on thermal status (see Chapter 6). Emotional stimuli, such as embarrassment or shock, can cause reflex changes in blood flow (such as facial blushing or the vasovagal response) as a result of activity of neurones in the cerebral cortex.

5.3 Origin of the stimulus for autonomic nervous activity affecting the cardiovascular responses to exercise

The cardiovascular responses to exercise match the intensity of work (figure 5.5). In the case of dynamic (isotonic) exercise, intensity is related to heart rate, cardiac output and oxygen consumption. In the case of static (isometric) exercise, intensity (active muscle mass and percentage of maximum voluntary contraction) is related to the increase in arterial blood pressure. In both cases sympathetic efferent activity is increased while parasympathetic efferent activity is decreased.

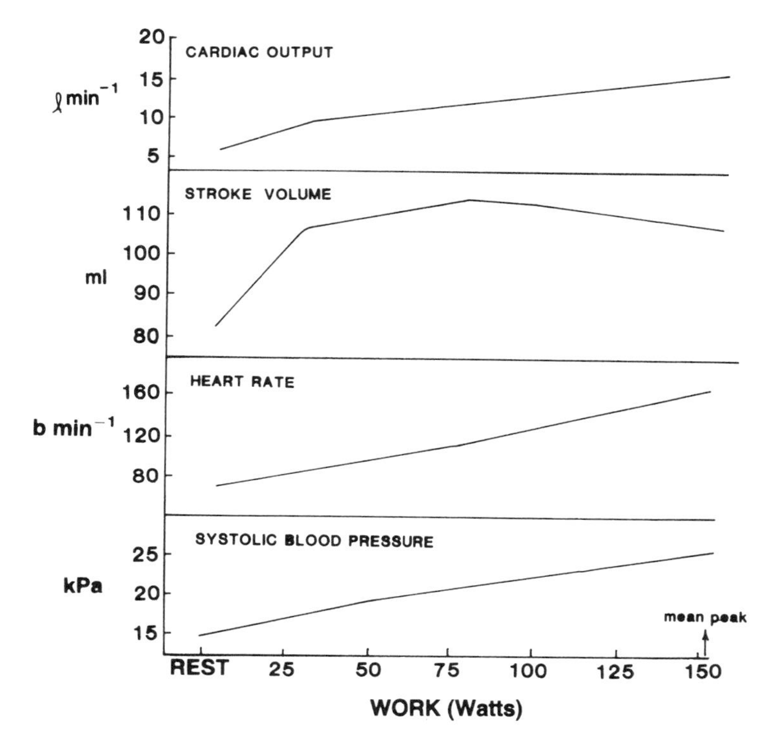

Figure 5.5 Mean haemodynamic responses to upright bicycle exercise in men aged 26–50 years. Data from Plotnick *et al.* (1986).

There are two autonomic neural mechanisms that operate, probably in concert, during exercise. One involves a reflex which originates from receptors located in the active skeletal muscles and joints, and the other involves impulses originating directly from higher centres.

Evidence for reflexes originating in the active muscles comes from studies in which the cardiovascular responses have been recorded in subjects whose leg muscles have been electrically stimulated thus eliminating any influence of central command. Such studies have shown that the cardiovascular responses to electrically stimulated muscular work, and the same work performed voluntarily, are the same. Further evidence for such a reflex mechanism comes from studies of animals in which the peripheral ends of the ventral roots of the motor nerve to a hind limb have been cut and stimulated electrically. This electrical stimulation leads to static muscle contraction and increases in arterial blood pressure, heart rate, left ventricular pressure and maximum development of left ventricular pressure. These cardiovascular responses are seen only when the dorsal roots are intact. Of the four groups of afferent fibres (see table 1.2) those in groups I and II do not mediate cardiovascular responses. The group III fibres are stimulated by the mechanical act of muscle contraction and fire vigorously at the onset of static muscular work. The group IV fibres also fire at the onset of muscular work but they seem to be stimulated by the metabolic response to muscular work and are especially active during static work associated with ischaemia. This 'metabolic reflex' allows changes in blood flow that are appropriate to the metabolic requirements of the tissues.

Evidence for central command comes from studies of the cardiovascular responses to dynamic and static exercise in subjects whose sensory afferent fibres arising from active skeletal muscle cells have been blocked by spinal anaesthesia. Such studies have shown that the cardiovascular responses to a given exercise are the same with or without spinal anaesthesia. Further evidence comes from studies in man and other primates using partial neuromuscular blockade, during which greater motor stimulation is required to achieve a given degree of muscle tension compared with the unblocked state. It has thus been demonstrated that the heart rate and blood pressure responses are greater during partial neuromuscular blockade, consistent with a central neural mechanism to explain the cardiovascular responses to muscular work.

A review of the neural reflex and central mechanisms that elicit cardiovascular responses to exercise is provided by Mitchell (1985).

5.4 Effect of autonomic nervous activity on the cardiovascular responses to exercise

5.4.1 HEART RATE

Before an individual begins to exercise, there is typically an anticipatory increase in heart rate. This is due to nerve impulses arising in the anterior aspect of the cerebral cortex and from diencephalic cells (in particular from the thalamus and hypothalamus) as well as adrenaline released from the adrenal medulla.

The increase in heart rate during exercise is proportional to the increase in work load. However, the absolute value of heart rate for a given amount of work is not constant. For example, a given work load in a high environmental temperature, or under emotional stress, causes a higher heart rate than the same work load performed in a cool environment, or in non-stressful conditions. Arm work is associated with higher heart rates at any given work load than leg work, and isometric exercise produces higher heart rates than isotonic exercise.

At the onset of exercise, heart rate increases so rapidly that it cannot be due to humoral factors. Despite the likelihood of a centrally mediated increase in heart rate in anticipation of exercise, there is evidence that the abrupt rise that occurs in the first 30–60 s of exercise is due solely to withdrawal of vagal tone since it can be blocked by atropine but not by β-adrenoceptor antagonists. It seems that the sympathetic nervous system contributes little to the increased heart rate at low work loads, even when the exercise is prolonged. Studies using infusions of atropine or propranolol (a non-selective β-adrenoceptor antagonist), double blockade (atropine + propranolol) or a combination of isometric and dynamic exercise suggest that only at maximal levels of exercise does removal of vagal tone no longer contribute to the increase in heart rate during exercise (figure 5.6). Thus at higher workloads the slower rise in heart rate that occurs between 30 and 240 s of exercise is due mainly to sympathetic nervous activity acting on the SA node. It seems feasible that whereas the rapid rise in heart rate during the first few seconds of exercise is brought about by a neural reflex mechanism involving muscle and joint receptors (group III fibre receptors) and central command, the slower rise in heart rate over the next 3 or 4 min is largely brought about by the activity of the chemoreceptors (group IV fibre receptors) in response to the metabolic changes in active muscles.

5.4.2 STROKE VOLUME

In a 70 kg individual, stroke volume is in the order of 70 ml, with 70 ml being ejected from the left and 70 ml being ejected from the right

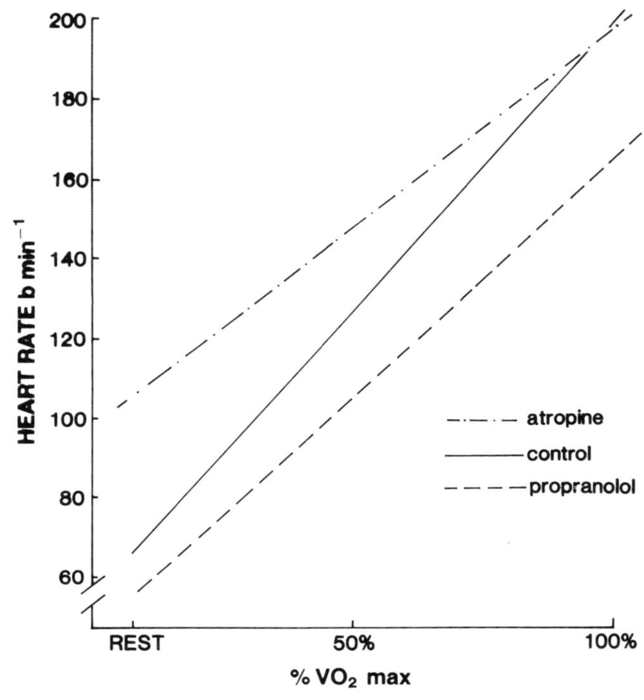

Figure 5.6 Effect of parasympathetic or sympathetic blockade on heart rate during exercise. Data from Ekblom *et al.* (1972).

ventricle virtually simultaneously each beat. Stroke volume is determined by:

1. the amount of stretch on the ventricular wall, which is related to the amount of blood entering the ventricle each beat, and
2. the degree of sympathetic nervous stimulation, which in turn determines the force of ventricular contraction.

These two factors are affected by two different physiological mechanisms, muscle fibre length (Frank–Starling mechanism) and sympathetic nervous activity.

Frank–Starling mechanism

The role of cardiac muscle cell length in determining stroke volume is expressed in the Frank–Starling law of the heart. This law states that the force of cardiac contraction is proportional to the initial length of the fibres. Thus, if cardiac filling is sufficient to stretch the cells, they will subsequently contract with greater strength. The Frank–Starling mechanism may not be independent of sympathoadrenal activity since cardiac filling may be augmented by catecholamine-mediated venoconstriction. Potentially the Frank–Starling mechanism could operate during exercise in view of the increased venous return that occurs. However, in man it seems to be of importance only at moderate levels of exercise before sympathetic nervous activity contributes greatly to cardiac contractility. Evidence for this is the failure of end-diastolic volume (i.e. cardiac filling) to increase beyond light exercise.

Sympathetic nervous activity

At higher work loads sympathetic nervous activity is of primary importance for the inotropic cardiac response. Since sympathetic nervous activity causes an increase in the force of cardiac contraction, such action is associated with a decrease in end-systolic volume (figure 5.7). In contrast with the dramatic rise in heart rate seen during maximal exercise (up to 300%), stroke volume increases very little (up to 30%). Unlike heart rate, stroke volume increases linearly with work up to only about 65–70% of the maximum work capacity. The reason for the plateau of stroke volume at high work loads is likely to be that the high heart rate dramatically reduces filling time. Since the amount of blood ejected is necessarily dependent on the blood that enters the heart, stroke volume will be compromised by reduced cardiac filling.

5.4.3 CARDIAC OUTPUT

Cardiac output is the amount of blood ejected from the heart each minute (that is, heart rate × stroke volume). For a resting individual this amounts to around 5.0 litres min^{-1} (72 b min^{-1} × 70 ml). Resting cardiac output is often expressed in relation to the body surface area as the **cardiac index**. Thus, the cardiac index of an individual with a cardiac output of 5.0 litres min^{-1} and a body surface area of 1.9 m^2 is 3.0 litres m^{-2} (5.0/1.9).

During exercise, despite the plateau of stroke volume, cardiac output increases linearly with work load to around 21 litres min^{-1} in

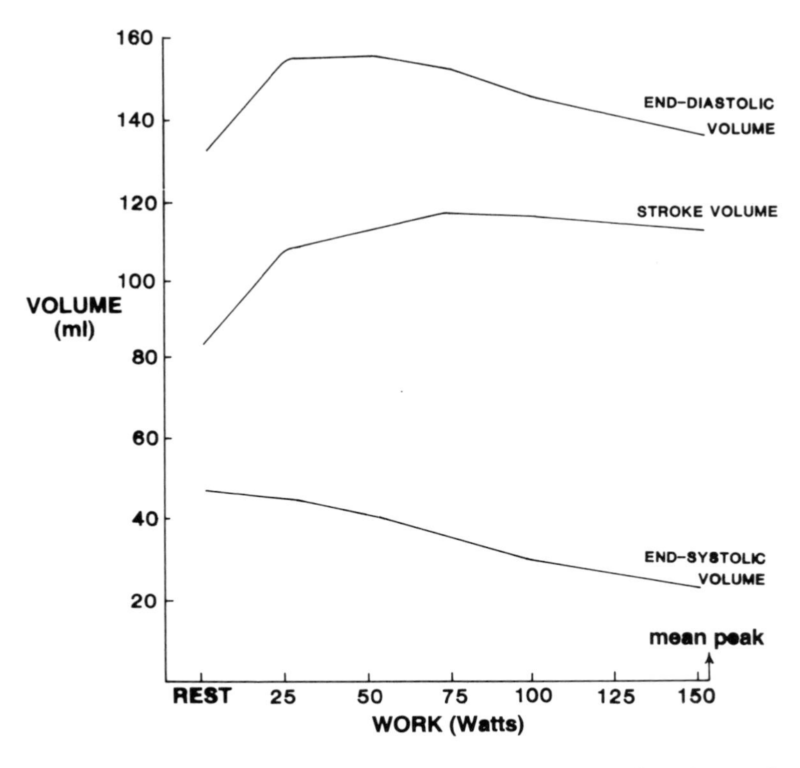

Figure 5.7 Stroke volume and end-diastolic and end-systolic volumes during upright bicycle exercise in men. Data from Plotnick *et al.* (1986).

the average 70 kg individual. The continued increase in cardiac output at maximal stroke volume is due to the increase in heart rate. The relationship between cardiac output and work load is of functional significance for blood flow and therefore oxygen delivery to, and metabolite removal from, the working muscle cells.

5.4.4 BLOOD FLOW

Skeletal muscle blood flow

Muscle blood flow may increase 25-fold during heavy exercise. During dynamic exercise more than 80% of all the muscle capillaries may be open. In contrast, during isometric exercise blood flow to the active muscle may be occluded by compression of the blood vessels due to the muscular contraction.

In order for muscle blood flow to increase during dynamic exercise, there must be a decrease in blood vessel resistance to flow, or an increase in driving pressure, or both. During isotonic exercise, blood flow to exercising muscle increases due to both an increase in cardiac output and a decrease in muscle vascular resistance.

Humoral factors affecting vascular resistance Humoral factors that affect vascular resistance include circulating catecholamines as well as other vasoactive hormones and local changes in pO_2, K^+, pCO_2, adenosine, pH and phosphate ions. The latter accumulate as a result of muscle contraction and lead to relaxation of the smooth muscle of the arterioles supplying skeletal muscle.

Systemic arterial blood pressure When the intensity of dynamic exercise exceeds about 80% of maximum oxygen consumption, the noradrenergic nervous stimulus is sufficiently great to bring about α-adrenoceptor-mediated vasoconstriction of the skeletal muscle arterioles. This vasoconstriction overrides the vasodilator effect of local metabolites and is a crucial means of maintaining arterial blood pressure.

During both isotonic and isometric exercise the baroreceptor response is set at a higher level. That is, the threshold of blood pressure at which the baroreceptors fire is higher. This is consistent with the fact that mean arterial blood pressure can be increased dramatically during vigorous isotonic or isometric exercise (especially in sedentary individuals), thus contributing to increasing blood flow to meet the metabolic demands of the tissues. The sensitivity of the arterial baroreceptors (and chemoreceptors) may be altered as a result of an increased level of K^+ in the extracellular fluid (as a result of muscular contraction) leading to a decrease in the cell membrane potential. Despite a resetting of the baroreceptors to a higher level, they are still able to regulate blood pressure, that is, the gain of the mechanism is not altered by exercise.

Redistribution of cardiac output In order to allow an increased proportion of cardiac output to be delivered to skeletal muscle, blood flow to inactive tissues is reduced (table 5.1). Thus, during maximal

Table 5.1 Distribution of cardiac output at rest and in maximal exercise. (Typical values for a healthy adult.)

Tissue	Rest ($ml\ min^{-1}$)	Maximal exercise ($ml\ min^{-1}$)
Heart	210	840
Active muscle	525	18 375
Inactive muscle	525	315
Brain	600	600
Skin	400	400
Rest of body (kidney, liver, GI tract, etc.)	2740	470
Cardiac output	5000	21 000

exercise, renal and splanchnic blood flow may be reduced to around 25% of the resting value.

Myocardial blood flow

Myocardial blood flow increases during exercise, primarily as a function of pO_2. Other humorally mediated vasodilators (CO_2, H^+, lactate and adenosine) have less potent effects on myocardial blood flow.

Skin blood flow

Skin blood flow decreases at the onset of exercise as a result of an increase in sympathoadrenal activity. After about 5 min of exercise, skin blood flow increases, possibly due to a neural vasodilatory mechanism or in association with bradykinin formed during sweat-gland activity.

Cerebral blood flow

It is unlikely that there is any change in cerebral blood flow during exercise since mean pCO_2, to which the cerebral vessels are exquisitely sensitive, is usually maintained during exercise.

5.4.5 VENOUS RETURN

Cardiac output is determined by venous return (the amount of blood returning to the heart each minute), and this is controlled by peripheral factors. Noradrenergically mediated vasoconstriction in inactive muscle and splanchnic and renal vascular beds (table 5.1) offsets the vasodilatation in active muscle, thus preventing a catastrophic decrease in total peripheral resistance during exercise which is essential to maintain blood pressure and therefore blood flow back to the heart. A decrease in total peripheral resistance during exercise is inevitable since mean arterial blood pressure is essentially unchanged while cardiac output increases. In a resting individual, total peripheral resistance is in the order of 2.48 peripheral resistance units (PRU) (12.4 kPa, mean arterial pressure (MAP)/5.0 litres min^{-1}, cardiac output ($\dot{Q}$)) and during maximal exercise decreases to around 0.76 PRU (16.0 kPa (MAP)/21.0 litres min^{-1}($\dot{Q}$)). The tissues that receive a reduced blood supply are able to continue to support oxidative metabolism since they are able to increase the extraction of oxygen per unit of blood.

The active muscles are a powerful aid to venous return. By rhythmically contracting around the veins they literally pump blood back to the heart, hence the term **muscle pump** used to describe this mechanism. Blood is prevented from draining back into the muscle by the valves in the veins.

The increased rate and depth of pulmonary ventilation associated with exercise cause a decrease in intrathoracic pressure and an increase in abdominal pressure. The decrease in intrathoracic pressure causes a fall in right atrial pressure, whereas the increase in intra-abdominal pressure causes an increase in peripheral venous pressure. Since blood flows from a region of high to low pressure, these pressure changes also aid venous return.

5.5 Recovery from exercise

At the end of a bout of exercise, heart rate and stroke volume (and thus cardiac output) return to baseline levels fairly quickly in recovery. This contrasts with the slower decrease in muscle blood flow during recovery from exercise since it takes time for the humoral factors which affect muscle blood flow to be removed from the muscles and circulation. Since muscle receives such a huge proportion of cardiac output during vigorous exercise, it is not uncommon for blood pressure to decrease below baseline levels afterwards. If exercise is stopped abruptly and the individual is upright, this may be associated with syncope (fainting) as a result of decreased cerebral blood flow from the abrupt decrease in arterial blood pressure.

B: RESPIRATION DURING EXERCISE

5.6 Pulmonary ventilation

5.6.1 NEURAL CONTROL OF BREATHING

Breathing involves a rhythmical augmentory discharge from neurones with cell bodies in the medulla causing excitation of the diaphragm via the phrenic nerves and excitation of the intercostal muscles via the intercostal nerves. The latter nerves have their origin in the cervical and thoracic spinal column and receive preganglionic connections from neurones with cell bodies in the brainstem and medulla. The medullary neurones also receive inputs from neurones in the pons and from stretch receptors in the lungs. It seems likely that the medullary neurones possess inherent rhythmicity which accounts for the rhythmical nature of breathing. When the inspiratory neurones fire, inspiration takes place and when this firing ceases expiration follows. There are three stages in the rhythmical control of passive breathing:

1. initial inspiratory drive
2. cessation of inspiratory drive
3. continued cessation of inspiratory drive during expiration.

It seems likely that the initial inspiratory drive occurs due to inherent automaticity of firing of the medullary inspiratory neurones. The breathing rhythm may be generated by cells of the nucleus of the solitary tract which has axons that descend the spinal cord. The 'off' trigger seems to be produced partly by the inherent automaticity of the inspiratory neurones, together with synaptic inputs to the inspiratory neurones from other neurones in the medulla, from neurones with cell bodies in the pons and from stretch receptors in the lungs. The stretch reflex (**Hering Breuer reflex**) mediates inhibition of inspiration via stretch receptors in the lungs. When these receptors are stimulated, the resultant neural signals are relayed via the vagus nerve to the medullary respiratory centre, leading to a reflex decrease in tidal volume (the amount of air ventilating the lungs with each breath), but a compensatory increase in breathing frequency. The Hering Breuer reflex is pronounced in babies and animals but is less prominent in adult man. For a review of the control of breathing, see Wyman (1977).

5.6.2 MUSCULAR INVOLVEMENT IN BREATHING

In resting man inspiration involves the diaphragm, external intercostal muscles and, in some individuals, the scaleni muscles, whereas expiration is passive. However, when pulmonary ventilation exceeds about 50 litres min^{-1}, additional muscle groups are active in inspiration. These are the sternocleidomastoid, trapezius and pectoralis muscles. At ventilation rates above about 30 litres min^{-1} expiration is no longer passive as it involves contraction of the anteroabdominal wall musculature and the internal intercostal muscles. Consequently, during exercise, when ventilation rates may increase to as much as 200 litres min^{-1} in some individuals, the work of breathing is increased.

5.6.3 PULMONARY VENTILATION DURING EXERCISE

Pulmonary ventilation ($\dot{V}$) is determined by the breathing frequency (f) and the tidal volume (TV).

$$\dot{V} = f \times \text{TV (litres min}^{-1}\text{)}$$

The rate of breathing is determined by the length of time between bursts of motor unit activity, and the depth of respiration is determined by the number of motor units firing and the frequency of discharge.

The lungs of a resting 70 kg individual are ventilated typically by about 6.5 litres of air per minute or 12 breaths, each of 540 ml. In maximal endurance exercise, breathing frequency increases by about fourfold to 48 breaths per minute and tidal volume increases by about fivefold to 2.7 litres per breath, giving an increase in pulmonary ventilation in the order of 20 times above the resting value, some 130 litres min^{-1}.

During the first few seconds of steady-state exercise, ventilation increases rapidly and this is followed by a slower increase over the next 2–4 min. The rapid phase is attributed to a neurogenic mechanism whereas the slower phase is consistent with a humoral mechanism. The neural signals could theoretically be derived from central neurones of the motor cortex and/or from peripheral muscle and joint receptors.

When exercise is stopped, there is an immediate decrease in ventilation followed by a slower exponential decrease to the resting value. The immediate decrease in ventilation is so rapid that it is attributed to a neural mechanism, whereas the slower phase is attributed to a humoral mechanism.

At low work loads ventilation increases linearly with work rate. However, a so-called 'breaking point' of ventilation occurs when the exercise intensity exceeds around 50–60% of the maximum work capacity, corresponding to a heart rate of about 150 b min^{-1} in young adults. Since this event is associated with the accumulation of lactate and H^+, the ventilatory breaking point has been attributed to stimulation of the chemoreceptors in the carotid body due to acidaemia which gives rise to increased elimination of carbon dioxide via the lungs. This has the effect of restoring acid–base balance (the balance between compounds which give up H^+ and those which give up OH^- ions in the body fluids).

$$H^+ + HCO_3^- \rightarrow H_2CO_3 \rightarrow CO_2 + H_2O$$

5.7.1 OXYGEN TRANSPORT

Approximately 97% of oxygen which leaves the lungs in arterial blood is transported to the tissues in combination with haemoglobin in the erythrocytes. Blood haemoglobin is in the range of 2.2–2.8 mmol $litre^{-1}$ (14–18 g 100 ml^{-1} blood) in men and 1.8–2.5 mmol $litre^{-1}$ (12–16 g 100 ml^{-1}) in women. Haemoglobin is formed by bone marrow cells in the precursors of red blood cells. It is made of a protein, globin, and haem, the pigment that gives blood its colour. Haem comprises four pyrrole rings, numbered I–IV, connected by four methine (=CH–) bridges in which the carbon atoms are termed α, β, γ and δ. Each haem molecule has a globin polypeptide which is linked to it by the free amino acid, histidine. In the middle of the pyrrole ring and connected to the nitrogen atoms is an iron atom (figure 5.8). The lower range of normal blood haemoglobin concentration cited above in women compared with men is due to iron loss in blood during menstruation.

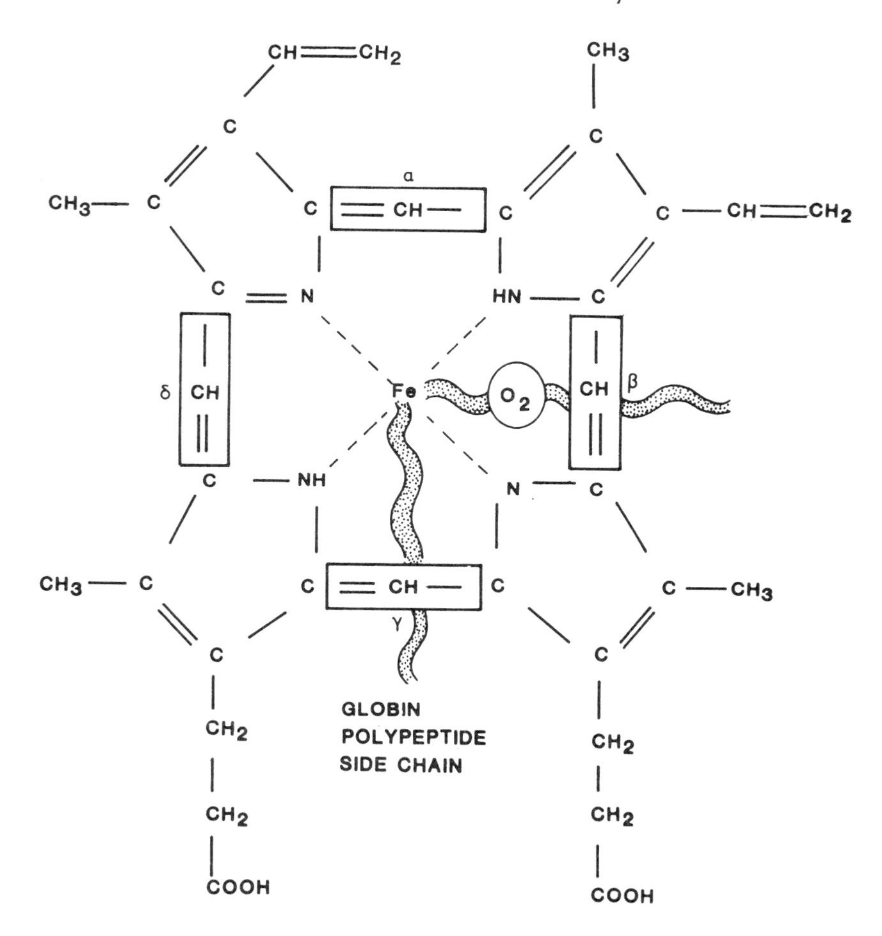

Figure 5.8 Structure of haemoglobin (other globin chains omitted for clarity).

Four of the haem molecules make one molecule of haemoglobin. Oxygen forms a loose association with haemoglobin by interrupting the globin side chain between the histidine molecule and the iron atom. The combination of oxygen with one molecule of haem increases the affinity of the subsequent haem molecule for oxygen. The successive enhancement of oxygen affinity for the next haem molecule accounts for the sigmoid ('S'-shaped) pattern of association or dissociation of oxygen with haemoglobin (figure 5.9).

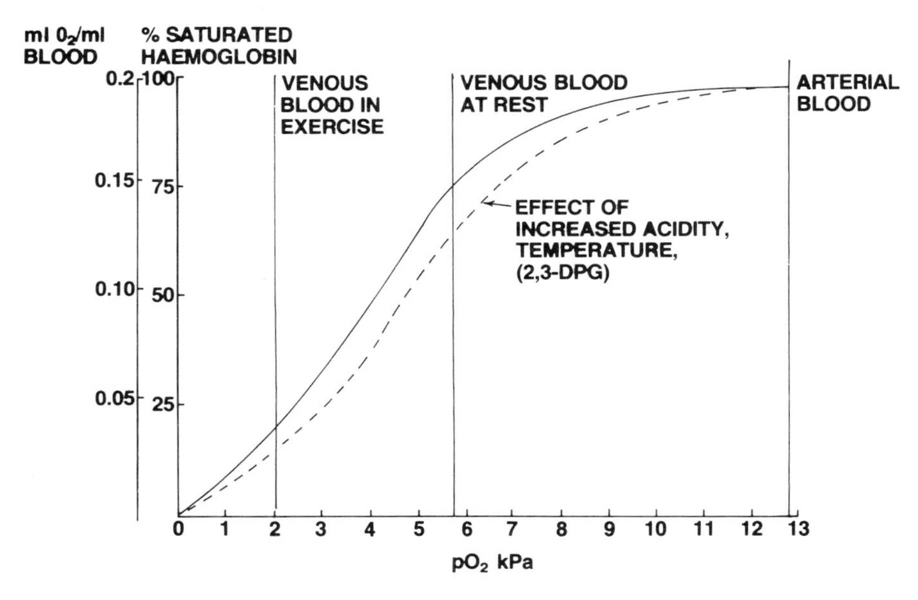

Figure 5.9 Oxyhaemoglobin dissociation curve. Extraction of oxygen by muscle.

Whilst most oxygen is transported in combination with haemoglobin, the remainder is dissolved in the plasma and intracellular fluid. Although very little oxygen is dissolved in this way (about 3% of total blood oxygen), it is this which is responsible for the blood or tissue oxygen tension (pO_2). Tension of gases in solution is equivalent to partial pressure in the gaseous phase. The difference in pO_2 across the cell membrane determines the flow of oxygen. When pO_2 in the tissue surrounding the capillaries is low, as in exercise, dissolved oxygen in the plasma moves towards the area of low pO_2. This low tissue pO_2 also facilitates the dissociation of oxygen from haemoglobin.

Oxygenated blood in the arteries leaving the lungs has a high pO_2 (12.9 kPa) and oxygen is fully bound to haemoglobin. As it travels through the capillary circulation, the blood pO_2 decreases to around 5.3 kPa but haemoglobin is still about 75% saturated with oxygen. The rate of oxygen use by cells is effectively controlled by the rate of ADP production which is a function of ATP hydrolysis. Thus, during exercise, when there is an increased production of ADP and thus increased oxygen use, tissue pO_2 may be reduced to about 2.0 kPa. Despite this relatively small drop in oxygen tension, the dissociation of oxygen from haemoglobin increases dramatically as a result of the facilitatory mechanism described earlier. In consequence during strenuous exercise haemoglobin in venous blood may be as little as 20% saturated with oxygen.

Three additional factors enhance oxygen release from haemoglobin during exercise. These are an increase in blood acidity, an increase in temperature and an increase in the concentration of 2,3-biphosphoglycerate (2,3-DPG; see page 132).
However, since patients with **McArdle's disease**, who lack muscle phosphorylase and are therefore unable to produce lactate and the associated increase in H^+, also display a breaking point in ventilation during exercise, this traditional explanation may be inadequate.

As the function of breathing is to deliver oxygen and remove carbon dioxide, one might expect that changes in pO_2, pCO_2 and/or pH (since $CO_2 + H_2O \rightarrow HCO_3^- + H^+$) would modify ventilation. The role of the carotid and aortic bodies and central chemoreceptors in detecting changes in these gas tensions was discussed in Chapter 1. However, as indicated in that chapter, mean arterial pO_2 and pCO_2 are unchanged during exercise. Yet, even when mean arterial pH is only slightly changed, there is an increase in ventilation. A possible explanation is that ventilation is controlled by the rate of carbon dioxide delivery to the lungs such that ventilation is controlled by breath-to-breath changes in pH. This would be consistent with carotid sinus function being flow dependent, and there is no reason why the carotid bodies are not affected by flow as well as pressure.

A further possibility is that since the arterial potassium ion concentration is temporally similar to the pattern of change of ventilation during exercise, an increase in the arterial concentration of this ion may stimulate the arterial chemoreceptors to bring about an increase in pulmonary ventilation.

5.7 Tissue respiration

Blood acidity

A decrease in blood pH and an increase in pCO_2 cause greater release of oxygen from haemoglobin in the erythrocytes. This effect is most pronounced during vigorous exercise, in association with the production of lactate in skeletal muscle, and permits around 10% greater release of oxygen from haemoglobin. This effect of acidity (the **Bohr effect**) shifts the oxygen/haemoglobin dissociation curve such that there is a higher pO_2 in the capillary blood at the same percentage saturation of haemoglobin with oxygen.

2,3-Biphosphoglycerate (2,3-DPG)

The abbreviation of 2,3-biphosphoglycerate to 2,3-DPG instead of 2,3-BPG is a relic of the old term 2,3-diphosphoglycerate. This molecule is formed in the red cells as a result of glycolysis. 2,3-biphosphoglycerate binds to deoxygenated haemoglobin and facilitates the release of oxygen from oxyhaemoglobin, thus shifting the oxygen/haemoglobin curve to the right.

Temperature

A rise in blood temperature also shifts the oxygen/haemoglobin curve to the right. In addition, the vasodilatation associated with metabolic heat production causes an increase in the capillary surface area for diffusion of oxygen to the tissues.

5.7.2 OXYGEN DELIVERY TO AND EXTRACTION BY SKELETAL MUSCLES

During exercise the diffusion capacity of oxygen from the alveoli into the blood increases as a result of an increase in the rate of pulmonary blood flow (since diffusion capacity is the volume of gas which diffuses through the alveolar capillary membrane per minute per pressure difference of 1 mm Hg or 0.13 kPa). Thus, even in strenuous work, blood leaving the lungs is virtually fully saturated with oxygen.

Oxygen delivery to muscle increases in exercise owing to an increase in blood flow, and in addition to an increased delivery of oxygen the muscle cells are able to extract more oxygen per unit of blood. The latter is brought about by an increased gradient of oxygen tension across the muscle cell membrane as intracellular pO_2 decreases with increasing muscle work. Oxygen extraction is assisted by the influence of local factors, primarily pH and temperature which in exercise shift the oxyhaemoglobin dissociation curve to the right (figure 5.9). Oxygen extraction is related to the mean transit time of the red cells through the open capillaries, with large extractions of oxygen occurring when the mean transit time (MTT) is slow. The increase in capillary density that occurs with training is associated with a slower MTT for a given rate of blood flow. Therefore, as the MTT of the red cells is slower, training is accompanied by improved oxygen extraction for any given rate of blood flow.

5.7.3 MYOGLOBIN

Myoglobin is found in skeletal muscle, especially in red slow twitch fibres and cardiac muscle cells. It extracts oxygen from the blood haemoglobin and in this way oxygen is transported into the muscle cells. Myoglobin thereby increases the amount of oxygen which enters the muscle cell, and acts as a temporary storage molecule for a small amount of oxygen.

5.7.4 CARBON DIOXIDE TRANSPORT IN BLOOD

Carbon dioxide diffuses from the tissues to the blood along a tension or concentration gradient. A small amount, around 7%, is transported dissolved in blood. The remainder diffuses into the erythrocytes. There about three-quarters of it forms carbonic acid by reacting with the cell water.

$$CO_2 + H_2O \xleftrightarrow{\text{carbonic anhydrase}} H_2CO_3$$

The carbonic acid ionizes to form hydrogen and bicarbonate ions.

$$H_2CO_3 \leftrightarrow H^+ + HCO_3^-$$

The hydrogen ions combine with haemoglobin in the erythrocytes while the bicarbonate diffuses out into the plasma.

The remaining 25% of carbon dioxide which enters the erythrocyte combines loosely with haemoglobin forming carbaminohaemoglobin, and the CO_2 is released later from the lungs during expiration. The affinity of haemoglobin for carbon dioxide or carbon monoxide is much greater than for oxygen, which explains why breathing high concentrations of either of these gases can quickly cause death.

A carbon dioxide dissociation curve is drawn in the same way as for oxygen. However, unlike the curve for oxygen the normal range of

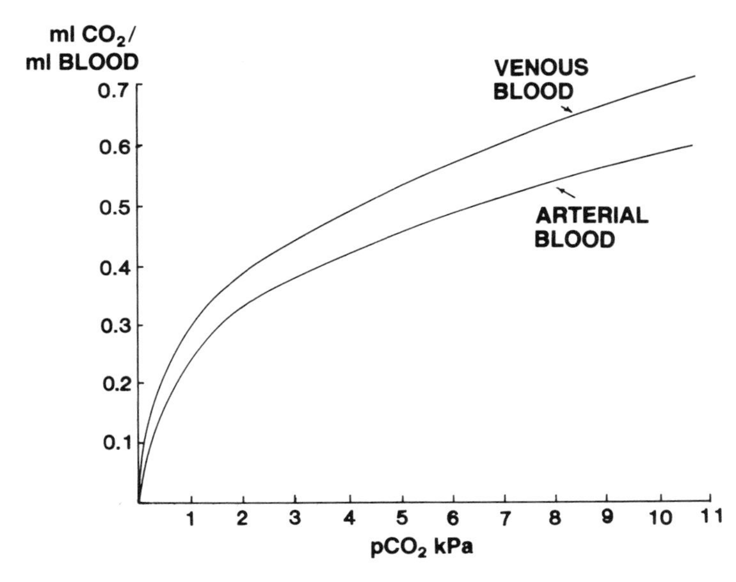

Figure 5.10 The carbon dioxide dissociation curve.

pCO_2 varies very little between venous and arterial blood, being 5.3 kPa in arterial blood and 6.0 kPa in venous blood (figure 5.10).

The formation of deoxygenated haemoglobin after oxygen has been delivered to the tissues thus provides an important mechanism for buffering H^+ and maintaining acid–base balance. The effectiveness of haemoglobin in transporting H^+ and CO_2 explains why blood pH decreases so little (by only about 0.04 of a pH unit), even in very vigorous exercise.

5.8 Oxygen consumption

Tissue oxygen consumption depends on the delivery of oxygen (arterial blood flow and arterial oxygen content) and the extraction and use of oxygen (arterial–venous difference for O_2) by the tissue in question.

Thus, $VO_2 = Q \times a - \bar{v}\ O_2$ difference where VO_2 is the volume of oxygen consumed, Q is blood flow, and $a - \bar{v}\ O_2$ is the difference in oxygen content between arterial and venous blood.

It is usually more meaningful to express both oxygen consumption and blood flow as rates per unit time ($\dot{V}O_2$ and $\dot{Q}$). In addition, these values may be expressed in relation to the weight of the tissue.

The factors controlling muscle blood flow and the $a - \bar{v}$ O_2 value have already been outlined in this chapter.

5.8.1 SKELETAL MUSCLE OXYGEN CONSUMPTION

Resting skeletal muscle blood flow is in the order of 3 ml blood 100 g muscle^{-1} min^{-1}. Each 100 ml arterial blood contains close to 19.5 ml oxygen (0.195 ml, or 8.7 μmol, oxygen per ml blood). Resting muscle extracts about 25% of this oxygen, giving an $a - \bar{v}$ O_2 value of 0.049 ml oxygen (2.2 μmol) per ml blood (0.25 × 0.195). Therefore:

$$\begin{aligned} \text{Resting muscle } \dot{V}O_2 &= 3 \times 0.049 \\ &= 0.15 \text{ ml } 100 \text{ g}^{-1} \text{ min}^{-1} \\ &\quad (6.7\ \mu\text{mol } 100 \text{ g}^{-1} \text{ min}^{-1}) \end{aligned}$$

For a standard 70 kg individual with 35 kg of muscle this amounts to 52.5 ml min^{-1}, about a fifth of total body oxygen consumption at rest (see page 136).

During maximal vigorous exercise, active muscle can receive up to 75 ml blood 100 g^{-1} min^{-1}. The oxygen content of arterial blood is the same at rest and during exercise. However, during maximal vigorous exercise, active muscle extracts about 75% of this oxygen. Thus the $a - \bar{v}$ O_2 value is 0.146 ml (6.5 μmol) oxygen per ml blood (0.75 × 0.195). Therefore:

$$\begin{aligned} \text{Exercising muscle } \dot{V}O_2 &= 75 \times 0.146 \\ &= 10.95 \text{ ml } 100 \text{ g}^{-1} \text{ min}^{-1} \\ &\quad (489\ \mu\text{mol } 100 \text{ g}^{-1} \text{ min}^{-1}) \end{aligned}$$

For the same individual using about 70% of his muscle mass (0.70 × 35 kg), this amounts to 2.68 min^{-1}, which is close to the maximum oxygen consumption of the total body (see page 136).

5.8.2 CARDIAC MUSCLE OXYGEN CONSUMPTION

Cardiac muscle oxygen consumption increases only through an increase in blood flow since, in this organ, oxygen extraction is virtually maximal even at rest. At rest coronary blood flow is in the order of 70 ml 100 g^{-1} min^{-1} and oxygen extraction is as high as 80%. Thus, the $a - \bar{v}$ O_2 value is 0.156 ml oxygen (7 μmol) per ml of blood (0.80 × 0.195). Therefore:

$$\text{Resting coronary } \dot{V}O_2 = 70 \times 0.156$$
$$= 10.92 \text{ ml } 100 \text{ g}^{-1} \text{ min}^{-1}$$
$$(490\ \mu\text{mol } 100 \text{ g}^{-1} \text{ min}^{-1})$$

For an individual with a 300 g heart this amounts to 32.76 ml min^{-1} (1.46 mmol min^{-1}), about 13% of total body oxygen consumption at rest. Coronary blood flow is autoregulated by local factors, and in this way keeps pace with metabolic demands for oxygen. During maximal vigorous exercise, coronary blood flow increases proportionally to the increase in cardiac work.

$$\text{Exercising coronary } \dot{V}O_2 = 280 \times 0.156$$
$$= 43.68 \text{ ml } 100 \text{ g}^{-1} \text{ min}^{-1}$$
$$(1.95 \text{ mmol } 100 \text{ g}^{-1} \text{ min}^{-1}),$$

or 131.04 ml min^{-1} (5.85 mmol min^{-1} for the whole heart)

Thus, on an equal weight basis cardiac muscle cells consume about 73 times as much oxygen in an individual at rest and four times as much during exercise as skeletal muscle cells. The much greater difference between cardiac and skeletal muscle oxygen consumption at rest compared with exercise is consistent with the fact that the cardiac muscle cells are much more active than skeletal muscle cells in a resting individual.

5.8.3 TOTAL BODY OXYGEN CONSUMPTION

Total body oxygen consumption is directly determined from cardiac output (total body blood flow) and the $a - \bar{v}$ O_2 value measured in the left and right atria. Assuming a cardiac output of 5.0 litres min^{-1} at rest and an oxygen extraction of 25% at rest, giving an $a - \bar{v}$ O_2 difference of 0.049 ml oxygen per ml blood (0.25 × 0.195):

$$\text{Resting oxygen consumption} = 5000 \times 0.049$$
$$= 245 \text{ ml min}^{-1}$$
$$(10.94 \text{ mmol min}^{-1})$$

This amounts to 3.5 ml kg^{-1} min^{-1} for a 70 kg individual.

Total body oxygen consumption during exercise

During exercise the increased oxygen consumption by skeletal and cardiac muscle cells is not balanced by a decreased oxygen consumption in other tissues. There is therefore an increase in total body oxygen consumption.

At the start of exercise, oxygen consumption increases sharply, reaching a plateau during steady-state exercise some 4–8 min later. Younger and fitter individuals display a plateau in oxygen consumption sooner than older and less fit individuals. Until this plateau is attained, the muscle cells rely on the anaerobic provision of ATP from stored phosphates and from anaerobic glycolysis (see Chapter 4).

Oxygen consumption increases linearly with work, except at levels approaching maximum when there is a relatively smaller increase in oxygen consumption for a given work increment. However, for the same amount of mechanical work, 10 min of weight lifting demands a higher oxygen consumption than treadmill running or cycling. In other words treadmill running and cycling are more efficient than weight lifting. The individual's perception of effort is related to the oxygen uptake rather than to the amount of mechanical work performed.

In dynamic exercise, the maximum rate of oxygen consumption is achieved when an increase in work load does not elicit any further increase in oxygen consumption. The maximum rate of oxygen consumption is therefore used as a laboratory measure of maximum aerobic capacity.

In maximal endurance exercise involving large muscle groups (such as running), cardiac output may increase up to about 21.0 litres min^{-1} and oxygen extraction increases to around 75%. Thus, the $a - \bar{v}\ O_2$ value increases to 14.62 ml oxygen (653 μmol) per ml blood (0.75×0.195). Therefore:

Maximum exercising oxygen consumption =

$$21\,000 \times 14.62 = 3.07 \text{ litres min}^{-1}$$

$$(134.24 \text{ mmol min}^{-1})$$

This is equivalent to 44 ml kg^{-1} min^{-1} (1.96 mmol kg^{-1} min^{-1}) for a 70 kg person.

From the preceding discussion it can be seen that the maximum amount of oxygen that can be consumed could be limited by one or both of the following factors:

1. oxygen extraction and use by the tissues
2. oxygen delivery to the tissues (blood flow).

Support for tissue oxygen extraction and use being a limiting factor has come from studies showing that breathing oxygen-enriched air failed to increase maximum oxygen consumption in exercising subjects, presumably because oxidative processes were already operating at their maximum capacity. Further support comes from studies in which individuals trained using one-legged cycling displayed higher peak oxygen consumptions during exercise with the trained leg compared with the untrained leg. Indeed, studies of mitochondrial enzyme activity with long-term (several months) training show that there are circulatory and metabolic explanations for why individuals can attain higher maximum rates of oxygen consumption after training.

However, the most likely explanation is that oxygen delivery in blood is the major limiting factor for maximum oxygen uptake. This view is substantiated by studies which have shown a close association between heart size and stroke volume with maximum oxygen consumption. It can be given further support by considering that the maximum amount of blood that the skeletal muscle vasculature can accommodate is around 2 litres $min^{-1} kg^{-1}$, and this amounts to a total of 60 litres min^{-1} in an adult man with 30 kg of muscle. The fact that cardiac output rarely exceeds half this amount in trained adult male athletes provides a strong argument that the upper limit of cardiac output is controlled by a cardiac mechanism.

Nevertheless, it is not clear whether blood flow (i.e. cardiac output) is limited by an inadequate driving pressure (mean arterial blood pressure) or by an inadequate peripheral resistance which would set a limitation on venous return. This problem is addressed in detail in a review by Rowell (1974).

Total body oxygen consumption during recovery from exercise

Following exercise there is a rapid decrease in oxygen consumption followed by a slower decline to baseline values. This **excess post-exercise oxygen consumption (EPOC)** (formerly given the misnomer 'oxygen debt') is used partly to replenish intramuscular phosphate as well as for hepatic gluconeogenesis. Other reasons for the EPOC include an increased circulating concentration of hormones which increase metabolic rate (catecholamines and thyroxine), an increase in deep body temperature, restoration of the body's oxygen stores as well as a continued elevation of oxygen consumption by tissues which remain active following exercise (such as the respiratory muscles and cardiac muscle). For a review of the factors which account for the EPOC, see Gaesser and Brooks (1984).

References and further reading

Adams, L., Garlick, J., Guz, A., Murphy, K. and Semple, S.J.G. (1984) Is the voluntary control of exercise in man necessary for the ventilatory response? *J. Physiol.*, **355**, 71–83.

Asmussen, E., Nielsen, M. and Wieth-Pedersen, G. (1943) On the regulation of the circulation during muscular work. *Acta Physiol. Scand.*, **6**, 353–358.

Band, D.M., Wolff, C.B. and Ward, J. (1980) Respiratory oscillations in arterial carbon dioxide tension as a control signal in exercise. *Nature,* **283**, 4–5.

Berne, R.M. and Levy, M.N. (1981) *Cardiovascular Physiology*, 4th edition, C.V. Mosby, St. Louis.

Blomqvist, C.G. and Saltin, B. (1983) Cardiovascular adaptations to physical training. *Ann. Rev. Physiol.*, **45**, 169–189

Ekblom, B., Goldbarg, A.N., Kilbom, A. and Åstrand, P.-O. (1972) Effects of atropine and propranolol on the oxygen transport system during exercise in man. *Scand. J. Clin. Lab. Invest.*, **30**, 35–42.

Freund, P.R., Rowell, L.B., Murphy, T.M., Hobbs, S.F. and Butler, S.H. (1979) Blockade of the pressor response to muscle ischaemia by sensory nerve block in man. *Am. J. Physiol.*, **237**, H433–H439.

Gaesser, G.A. and Brooks, G.A. (1984) Metabolic bases of excess post-exercise oxygen consumption: a review. *Med. Sci. Sports Ex.*, **16**, 29–43.

Gallo, L., Jr., Maciel, B.C., Marin-Neto, J.A., Martins, L.E.B., Lima-Filho, E.C. and Manço, J.C. (1988) The use of isometric exercise as a means of evaluating the parasympathetic contribution to the tachycardia induced by dynamic exercise in normal man. *Pflügers Arch.*, **412**, 128–132.

Gollnick, P.O., Armstrong, R.B., Saubert, C.W., Piehl, K. and Saltin, B. (1972) Enzyme activity and fiber composition in skeletal muscle of untrained and trained men. *J. Appl. Physiol.*, **33**, 312–319.

Hagberg, J., Coyle, E.M., Carroll, J.E., Miller, J.M., Martin, W.H. and

Brooke, M.H. (1982) Exercise hyperventilation in patients with McArdle's disease. *J. Appl. Physiol.,* **52**, 991–994.

Hultman, E. and Sjöholm, H. (1982) Blood pressure and heart rate response to voluntary and non-voluntary static exercise in man. *Acta Physiol. Scand.,* **115**, 449–501.

Kaijser, L. (1970) Limiting factors for aerobic muscle performance. *Acta Physiol. Scand. Suppl.,* **346**, 1–96.

Keele, C.A., Neil, E. and Joels, N. (1982) *Samson Wright's Applied Physiology*, 13th edition, Oxford University Press, Oxford.

Leonard, B., Mitchell, J.H., Mizuno, M., Rube, N., Saltin, B. and Secher, N.H. (1985) Partial neuromuscular blockade and cardiovascular responses to static exercise in man. *J. Physiol.,* **359**, 365–379.

Levy, M.N. (1984) Cardiac sympathetic–parasympathetic interactions. *Fed. Proc.,* **43**, 2592–2602.

Linton, R.A.F., Lim, M., Wolffe, C.B., Wilmhurst, P. and Band, D.M. (1984) Arterial plasma potassium measured continuously during exercise in man. *Clin. Sci.,* **67**, 427–431.

Maciel, B.C., Gallo, L., Neto, J.E.M., Filho, E.C.L. and Martins, L.E.B. (1986) Autonomic nervous control of the heart during dynamic exercise in normal man. *Clin. Sci.,* **71**, 457–460.

Mitchell, J.H. (1985) Cardiovascular control during exercise: central and reflex neural mechanisms. *Am. J. Cardiol.,* **55**, 34D–41D.

Orizio, C., Perini, R., Commande, A., Castellano, M., Beschi, M. and Veicsteinas, A. (1988) Plasma catecholamines and heart rate at the beginning of muscular exercise in man. *Eur. J. Appl. Physiol.,* **57**, 644–651.

Plotnick, G.D., Becker, L.C., Fisher, M.L., Gerstenblith, G., Renlund, D.G., Fleg, J.L., Weisfeldt, M.L. and Lakatta, E.G. (1986) Use of the Frank–Starling mechanism during submaximal versus maximal upright exercise. *Am. J. Physiol.,* **251**, H1101–H1105.

Rowell, L.B. (1974) Human cardiovascular adjustments to exercise and thermal stress. *Physiol. Rev.,* **54**, 75–159.

Rowell, L.B., Freund, P.R. and Hobbs, S.F. (1981) Cardiovascular responses to muscle ischemia in humans. *Circ. Res.,* **48**, Suppl. 1, 37–47.

Saltin, B. (1985) Hemodynamic adaptations to exercise. *Am. J. Cardiol.,* **55**, 42D–47D.

Saltin, B. (1988) Capacity of blood flow delivery to exercising skeletal muscle. *Am. J. Cardiol.,* **62**, 30E–35E.

Saltin, B., Nakar, K., Costill, D.L., Stein, E., Jansson, E., Essen, B. and Gollnick, P.D. (1976) The nature of the training response: peripheral and central adaptations to one-legged exercise. *Acta Physiol. Scand.,* **96**, 289–305.

Shepherd, J.T. and Mancia, G. (1986) Reflex control of the human cardiovascular system. *Rev. Physiol. Biochem. Pharmacol.,* **105**, 1–99.

Suarez, V.H., Messerti, F.H., Ventura, H.O., Aristimuno, G., Dresbenki, G.R. and Frohlich, E.D. (1982) Baroreceptor stimulation and isometric exercise in normotensive and borderline hypertensive subjects. *Clin. Sci.,* **62**, 307–309.

Victor, R.G., Seals, D.R. and Mark, A.L. (1987) Differential control of heart rate and sympathetic nerve activity during dynamic exercise. *J. Clin. Invest.,* **79**, 508–516.

Wyman, R.J. (1977) Neural generation of the breathing rhythm. *Ann. Rev. Physiol.,* **39**, 417–448.

Chapter 6

Temperature regulation

In Chapter 4 it was shown that most of the total energy from the breakdown of carbohydrate or lipid is transferred to ATP, with the remaining energy appearing as heat. During exercise the increased heat production in metabolism is lost by increasing skin blood flow and by the evaporation of sweat. Thermoregulation during exercise is discussed in the present chapter, together with the role of the autonomic nervous system in these responses.

6.1 Body temperature

Man has a core of tissues including the brain, heart, lungs, liver, kidneys and gastrointestinal structures, which are maintained close to 37°C. This core of warm tissues is surrounded by a shell of cooler subcutaneous tissues and the skin. The skin has no uniform temperature, with regional variation resulting from insulation provided by subcutaneous adipose tissue, vascularization and the metabolic activity of the underlying tissues. In some skin areas, particularly in the extremities, the temperature varies with changes in environmental temperature. In warm conditions the skin temperature over the whole body is more uniform than in cold conditions. This variability in skin temperature contrasts with the relative stability of the core temperature. There is a normal circadian variation in core temperature, with the lowest levels being recorded at around 05.00 h and the highest levels at around 13.00 h. Over a 24-h day/night cycle the maximum variation in core temperature is between 0.2 and 1°C.

Body core temperature increases when the heat gained by the body exceeds the heat lost. This may be due to the heat loss mechanisms failing to keep pace with cellular heat production (such as during exercise) or to environmental factors which are unfavourable to heat loss even though heat production may be at basal levels (such as long-term exposure to hot, humid conditions). The physical nature of heat transfer is outlined below.

6.2 Heat balance

6.2.1 HEAT GAIN

Heat is gained by the body through cellular metabolism (see Chapter 4). Metabolically produced heat is augmented at times of increased fuel metabolism such as during exercise or shivering.

The specific heat of the body tissues is 3.473 kJ kg^{-1} $°C^{-1}$. Thus for core temperature to increase by 1°C in a 70 kg individual there must be a net gain in heat of 243 kJ. Heat production is proportional to oxygen consumption.

During exercise the increase in energy expenditure would quickly cause a fatal increase in heat gain if it were not for the efficiency of the heat loss mechanisms. This is illustrated in the following example concerning a hypothetical marathon runner. This individual weighs 70 kg, has a maximum oxygen consumption of 5.0 litres min^{-1} and completes a marathon race in 2.5 h. He runs the race at 70% of his maximum oxygen consumption, which means a total consumption of 525 litres of oxygen (0.70 × 5.0 litres min^{-1} × 150 min). During the run he has derived equal amounts of ATP from carbohydrate and lipid metabolism, and therefore 20.2 kJ of energy have been expended per litre of oxygen consumed (table 4.4), giving a total heat gain of 151.5 kJ kg^{-1} (525 litres × 20.2 kJ $litres^{-1}$/70 kg). Since the specific heat of the body tissues is 3.473 kJ, in the absence of any loss this heat gain would be sufficient to raise the mean body temperature by 43.6°C (151.5/3.473). The fact that mean body temperature increases by only around 4°C during a marathon run (close to the upper limit for survival) is indicative of the efficiency of the heat loss mechanisms.

6.2.2 HEAT EXCHANGE

Energy in the form of heat is exchanged between two objects across a thermal gradient, with the hotter object losing heat to the cooler object. This observation is central to the second law of thermodynamics. The spontaneous ways in which heat is exchanged between the human body and the environment are by radiation, convection and conduction.

Radiation

Emission of radiant, or electromagnetic, energy is expressed as: $R = \varepsilon.\sigma.A.T^4$, where:

R = rate of radiant heat exchange in watts
ε = the emissivity of the surface
σ = the Stefan–Boltzmann constant (5.67×10^{-8} W m^2 K^4)
A = the surface area in m^2
T = the surface temperature in Kelvins.

Emissivity varies between 0 for a perfect reflector such as a mirror and 1.0 for a perfect black body which absorbs all the radiant energy it receives. Since skin is considered to behave as a black body, its emissivity is close to 1.0. Thus with little error it can be assumed that radiant heat loss is proportional to the skin temperature raised to the fourth power.

Convection

Convection involves heat exchange with the air. The amount of heat lost depends on the thermal gradient between the skin (or outer layer of clothing) and the air, as well as the amount of air movement (wind chill). Convection is proportional to the square of air velocity. Clearly if the air temperature is lower than skin temperature, which is usually the case, heat will be lost by the body.

Conduction

Heat exchange by conduction through contact with solid matter usually amounts to a negligible heat loss since so little of the body is exposed to material which conducts heat well, and the conductivity of air is very low (0.025 W m^{-1} °C). Under most circumstances clothing is a good insulator. However, if clothing is wet, its conductance increases and body heat loss increases. Similarly, if an individual is submerged in water, heat loss may be quite considerable since the thermal conductivity of water is quite high (0.60 Wm^{-1} °C).

6.2.3 HEAT LOSS

Evaporation

Heat is lost by evaporation through the conversion of liquid water to water vapour. In this process, 2.43 kJ of heat are removed per 1 ml water evaporated at physiological temperature. Both sweat and expired air contribute to the water that is evaporated, but in man the former is the more important.

Work

The energy in ATP from fuel metabolism is transferred to external work when exercise is carried out. Since man's mechanical efficiency is never more than about 25%, at least 75% of the total energy produced in metabolism to perform mechanical work is dissipated as heat.

Radiation, convection, conduction and evaporation are influenced by physical factors, such as the environmental conditions and clothing, as well as by physiological variables such as adiposity and cutaneous blood flow. For example, with respect to the environment, in a dry bulb temperature of 25°C and a relative humidity of 50%, radiation accounts for about 65% of heat loss, evaporation of sweat for about 25% and convection for the remaining 10%. The contribution of conduction is negligible. In a thermoneutral environment (28–30°C for a naked man) the contribution of radiation is less (about 40%) and convection more (about 30%) with evaporation of sweat accounting for slightly more heat loss than at 25°C. Again the contribution of conduction is negligible. In warm conditions the evaporation of sweat may contribute up to 90% of heat loss. The ability to lose heat by sweating is crucially dependent on the humidity of the air. An increase in air humidity reduces its water-carrying capacity, thus reducing the amount of sweat evaporated. During exercise more sweat may be produced than is evaporated (depending on the intensity of exercise, environmental conditions, state of training and acclimation to the environmental conditions). In this case sweat is seen on the skin surface but unless it evaporates it has no effect on heat loss by evaporation. However, since it causes a reduction in skin temperature it has the effect of altering the temperature gradient between the skin and environment, which has implications for heat exchange by radiation and convection. By wetting the clothes, unevaporated sweat also potentially modifies heat loss by conduction.

An important physiological contribution to heat transfer is **countercurrent heat exchange**. This term describes the exchange of heat between arterial and venous blood. The mechanism is particularly effective in the limbs and digits of man and in the carotid rete of the dog. In a cold environment blood flow through the arteriovenous anastamoses and superficial veins is reduced by vasoconstriction. Thus, the heat that is transferred by conduction from the warm arterial blood to the cooler venous blood is not lost in great quantities to the environment. However, in a warm environment, blood flow through the arteriovenous anastomoses and superficial veins is high due to abolition of sympathetic vasoconstrictor tone, thus facilitating heat loss.

6.2.4 THE HEAT BALANCE EQUATION

From the above consideration of heat gain, heat exchange and heat loss, the heat balance equation can be derived.

$$S = M - E \pm R \pm C \pm K - W$$

where S = stored heat, M =metabolic heat production, E = evaporation, R = radiation, C = convection, K = conduction, W = work.

The components of the heat balance equation are conventionally expressed in kilojoules and related to body surface area. The units are therefore kJ m^{-2} h^{-1}. A hypothetical example will be discussed to illustrate the usefulness of the equation in determining heat transfer during exercise.

The following exercise study is carried out on a naked individual in an ambient (air and walls) dry bulb temperature of 25°C. The subject weighs 70 kg, is 1.80 m tall and has a surface area of 1.9 m^2. At rest the subject has a core temperature of 36.9°C. He then cycles at an even pace at a mechanical workload of 5.29 kJ min^{-1} (540 kg min^{-1}) for 1 h. During the exercise he attains a steady-state oxygen consumption of 1.25 litres min^{-1} (55.80 mmol min^{-1}), a ventilation rate of 30 litres min^{-1} and a non-protein respiratory exchange ratio of 0.86. He sweats, but not profusely enough for the sweat to drip off the skin. Immediately on finishing the exercise his weight has decreased to 69.65 kg, indicating that 350 g has been lost by evaporation of sweat and water from the respiratory tract. After towelling himself dry the subject is reweighed and his weight is now 69.60 kg indicating that a further 50 g of sweat was produced but did not evaporate and therefore did not contribute to heat loss. During the cycle, core temperature increased by 0.7°C to 37.6°C. Mean skin temperature decreased from 31.0°C at rest to 29.7°C by the end of the exercise, averaging 30°C.

Average body temperature

The heat balance equation requires measurements of average body temperature, derived from measurements of core and mean skin temperature. Two formulae which have been recommended for man are:

$$(0.65 \times \text{core}) + (0.35 \times \text{skin}) \text{ (Burton, 1935)}$$
$$(0.80 \times \text{core}) + (0.20 \times \text{skin}) \text{ (Hardy and DuBois, 1938)}$$

The choice of formula depends on the environmental temperature, with the core to skin ratio varying between 0.5:0.5 and 0.8:0.2 for environmental temperatures between 22 and 36°C, as well as the length of time the individual is exposed to a given environmental temperature.

Using Burton's formula in the present example (since this is the more appropriate for the environmental temperature), the average body temperature was 34.83°C at the start of the cycle $(0.65 \times 36.9) + (0.35 \times 31.00)$ and increased by 0.5°C to 35.33°C $[(0.65 \times 37.6) + (0.35 \times 29.7)]$.

It is now possible to calculate the heat balance equation.

Stored heat

Stored heat is calculated from the change in average body temperature (0.5°C) and the specific heat of the tissues (3.473 kJ kg^{-1} °C):

$$0.5 \times (3.473 \times 69.60) = 120.86 \text{ kJ h}^{-1}$$

Correcting for surface area the stored heat in this example is *63.61 kJ m^{-2} h^{-1}.*

Metabolic heat

Metabolic heat production can be calculated from the heat equivalent of an oxygen consumption of 75 litres (1.25 litres min^{-1}) for a non-protein respiratory exchange ratio of 0.86 using standard values of energy equivalents (see table 4.4). In this case 20.3 kJ of heat are produced per litre of oxygen consumed.

$$75 \times 20.3 = 1.522 \text{ MJ h}^{-1}$$

Correcting for surface area the metabolic heat production is *801.31 kJ* $m^{-2}\ h^{-1}$.

Heat loss by radiation

Radiant heat loss is calculated as the difference between emitted radiation and absorbed radiation.

$$\text{Emitted radiation} = R_E = \varepsilon.\sigma.A.T^4$$
$$\text{Absorbed radiation} = R_A = \varepsilon.\sigma.A.T^4$$

$$\text{Therefore, net heat loss by radiation} = R_E - R_A = R$$
$$R = \varepsilon.\sigma.\alpha\,(T^4{}_E - T^4{}_A)$$

Thus, in this example, T_E is the mean skin temperature, which is 30°C (303 K) and T_A is the ambient temperature, which is 25°C (298 K). It is assumed that the emissivity of the skin is 0.99.

$$\text{Thus, } R = (0.99)\,(5.67 \times 10^{-8}\text{ W m}^2\text{ K}^4)\,(1.9\text{ m}^2)\,(303–298\text{ K})$$
$$= 53.33\text{ W}$$

Converting watts (J s^{-1}) into kJ h^{-1} and correcting for surface area, heat loss by radiation is *101.05 kJ* $m^{-2}h^{-1}$.

Heat loss by evaporation

The loss of 350 g of water includes loss from sweat and from the respiratory tract. Water loss from the respiratory tract will be considered first. The water content of fully saturated air at 101.3 kPa (760 mm Hg) and 30°C is 30 mg of water per litre of air. Assuming expired air is fully saturated and is a few degrees lower than core temperature as it leaves the mouth, a volume of 1800 litres (30 litres min^{-1}) expired during an hour of cycling would contain:

$$1800 \times 0.03 = 54\text{ g } H_2O$$

Assuming each millilitre of this water removes 2.43 kJ of heat, the heat loss would be:

$$54 \times 2.43 = 131.22 \text{ kJ h}^{-1}$$

Correcting for body surface area this becomes *69.06* $kJ\ m^{-2}h^{-1}$.

Thus sweat evaporation amounts to 350–54 g, which is 296 g. Therefore, heat loss from sweat evaporation is:

$$296 \times 2.43 = 719.28 \text{ kJ h}^{-1}$$

Correcting for body surface area this is equivalent to *378.57* $kJ\ m^{-2}\ h^{-1}$.

Heat lost through mechanical work

The individual carries out 317.40 kJ of work in the hour he is cycling (5.29 kJ min^{-1}), which corrected for body surface area amounts to a transfer of energy in ATP to work of *167.05* $kJ\ m^{-2}\ h^{-1}$.

Thus far, then:

$$\begin{array}{ccccccccc} S & = & M & - & E & - & W & - & R \pm C \pm K \\ 63.61 & = & 801.31 & - & (378.57 + 69.06) & - & 167.05 & - & 101.05 \pm C \pm K \end{array}$$

Since the heat exchange by convection and conduction cannot be calculated separately, they must be considered together. However, since heat loss by conduction is negligible, most of the remaining heat loss can actually be attributed to convection.

Therefore, rearranging the heat balance equation:

$$M - E - W - S - R - (C + K) = 0$$

Therefore $(C + K) = -21.97\ kJ\ m^{-2}\ h^{-1}$

Since the value for convective and conductive heat exchange is negative, there must have been a net heat loss by these means. In summary, then, the overall heat exchange in this example can be written:

Heat production	−	heat storage	=	Heat loss				
				Evaporation		+ Work +	R	+ $(C + K)$
				sw	rt			
801.31	−	63.61	=	378.57	+ 69.06	+ 167.05	+ 101.05	+ (21.97)
				51%	9%	23%	14%	3%

where sw = sweat evaporation, and rt = evaporation from the respiratory tract.

6.3 Measurement of body temperature

6.3.1 CORE TEMPERATURE

Several sites are commonly used for the measurement of core temperature. These include the rectum, tympanic membrane, external auditory meatus, oesophagus and chest wall. Most methods involve the use of thermistors arranged in a conventional Wheatstone bridge circuit. Some of the advantages and disadvantages of the commonly used sites are discussed below.

Rectal temperature

Rectal temperature is measured by inserting a thermistor some distance (typically about 10 cm) past the external anal sphincter. The measurement is affected by the tissue mass, blood flow and the temperature gradient along the anal canal. Rectal temperature changes slowly compared with oesophageal, tympanic or oral temperatures due to the mass of tissue around the probe. Local blood flow also affects rectal temperature as evidenced by decreases in the temperature on standing from a seated position, or during the first few minutes of exercise. The temperature gradient along the anal canal may make as much as 0.8°C difference in the reading of rectal temperature. For this reason it is often preferable to report changes in rectal temperature rather than absolute values.

External auditory meatal temperature

The **external auditory meatus** is the canal which leads from the pinna of the outer ear to the tympanic membrane. It is about 2.5 cm long. Two branches of the external carotid artery, the maxillary and temporal arteries, pass close to the external auditory meatus. The measurement of temperature in the external auditory meatus overcomes the problems of discomfort and potential damage to the ear drum associated with tympanic thermometry. Again the thermistor must be insulated to ensure that the measurement is not influenced by environmental temperature. It may take up to 30 min for the temperature in the external auditory meatus to stabilize after insertion of the thermistor and insulation. The absolute measurement of external auditory meatal temperature is affected by a temperature gradient along the canal, and thus, like rectal temperature, changes in temperature are more reliable. Compared with rectal temperature, external auditory meatal temperature changes more quickly during exercise and displays an earlier plateau (figure 6.1). In this respect external auditory meatal temperature is similar to cranial temperature.

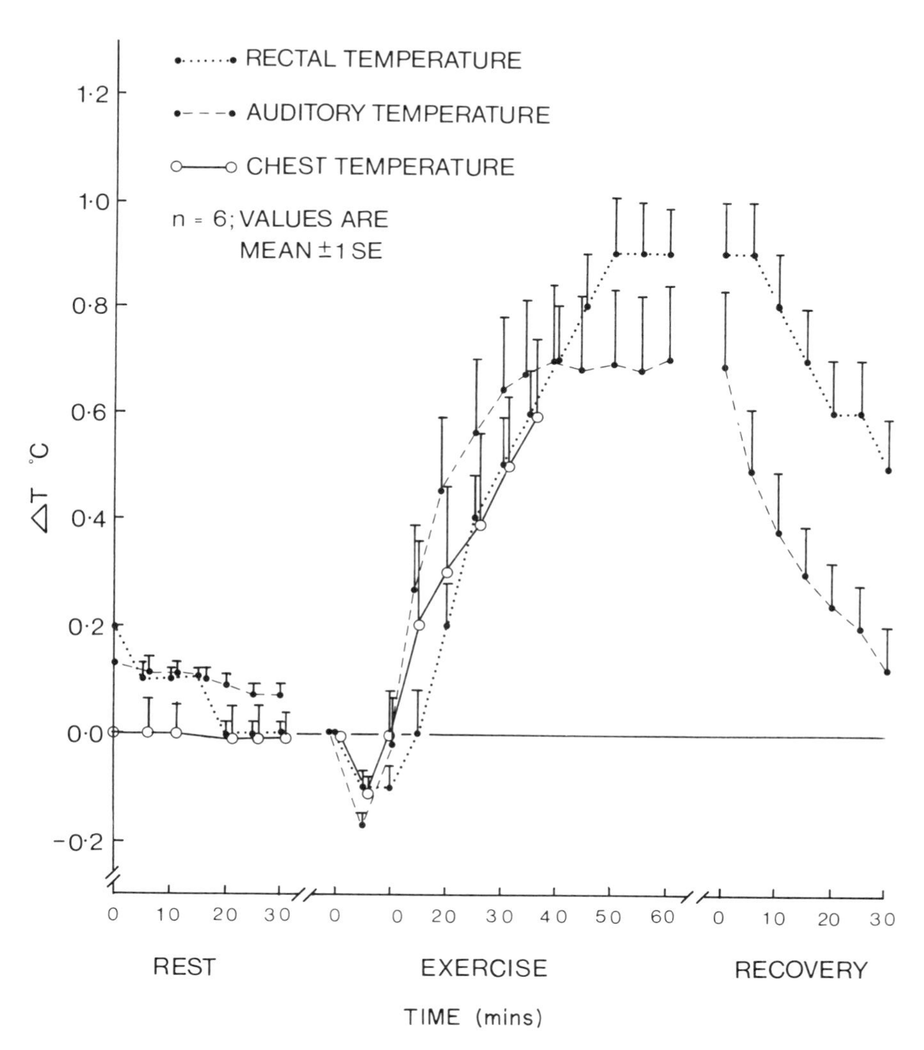

Figure 6.1 Change in rectal, external auditory meatal and chest wall temperature during moderate exercise. Unpublished data, Green and Cable (1986).

Oesophageal temperature

Oesophageal temperature is impractical for many situations, including exercise, since it changes due to swallowing and breathing even if the mouth is closed. There is a temperature gradient down the oesophagus and so, like rectal and external auditory meatal temperatures, changes rather than absolute temperatures should be used.

Chest wall

A relatively new method for measuring core temperature uses a flat probe taped to the skin of the chest (Fox *et al.*, 1973). The probe is heated to match deep body temperature so that it produces an area of zero heat flow across the underlying tissues. The probe comprises two thermistors which form two arms of a Wheatstone bridge circuit; one registers the temperature of the probe and the second that of the skin. Any change in skin temperature (due to heat flow from the core) relative to the probe produces an out-of-balance signal from the bridge which is amplified and digitally displayed. The method compares well with external auditory meatal temperature during exercise (figure 6.1) as well as with measurements of core temperature using a temperature-sensitive radio pill during changes in ambient temperature or injection of endogenous pyrogen to produce fever. However, the absolute temperature is lowered by excessive air movement. Under such conditions the changes in chest-wall deep body temperature are still reliable. A major limitation of this method during exercise is the effect of sweat on the adhesive material such that the probe tends to become detached from the skin when the skin is moist. It was for this reason that core temperature measurements were not obtained after 35 min of exercise in figure 6.1.

6.3.2 SKIN TEMPERATURE

Traditional methods of measuring skin temperature involve taping thermistors to specific skin sites. This requires small probes which do not insulate the skin and porous tape which allows air flow over the skin and the opportunity for sweat to evaporate freely.

Mean skin temperature

The skin covers a large surface area, around 1.9 m^2 in the average man, from which a mean is to be derived. The solution has been to weight the measurements from a discrete number of sites according to the surface area each represents. A number of weighting systems have been developed. Comparison of some of these different systems yields similar results.

The recent advent of infrared thermography has allowed the measurement of regional changes in skin temperature to be far more accurate (since the skin is not in contact with the measuring device) and more detailed (since the skin temperature of the whole body can be measured). Comparison between the conventional and infrared thermography methods in subjects running out of doors shows agreement of mean skin temperature to within 1.5°C.

6.3.3 AVERAGE BODY TEMPERATURE

The use of average body temperature has been advocated as an appropriate index of thermal input to the thermoregulatory centres because both central and peripheral thermoreceptors are involved in thermoregulation. However, this is of dubious value since mean skin temperature is weighted according to surface area rather than according to the distribution of cutaneous thermoreceptors. For example, there is a dense supply of thermoreceptors in facial skin and yet many of the weighted formulae for mean skin temperature do not include face skin temperature.

6.4 Physiological control mechanisms

Thermoregulation involves afferent input to various cells in the central nervous system. These inputs are integrated and result in a given autonomic response. Afferent impulses are processed in the spinal cord, the trigeminal nucleus, the thalamus, midbrain and hypothalamus. The dorsal horn cells of the spinal cord display exquisite sensitivity to incoming afferent signals from skin thermoreceptors. The trigeminal nucleus receives neural information from cold thermoreceptors in the face (demonstrated in the cat, rat and rabbit) which is then passed on to the thalamus. The thalamus also receives input from temperature receptors in the scrotum (as shown in the rat) or udder (demonstrated in the goat). The midbrain contains warm receptors which are sensitive to its own temperature but also detect changes in abdominal temperature (as shown in the rat). The midbrain may be part of a chain of cells which relays thermal information through the midbrain to the thalamus and finally to the hypothalamus. The hypothalamus receives an abundance of signals from the peripheral and central thermal neurones and has therefore been assumed to be the thermoregulatory centre. A further reason for assuming that the hypothalamus is the thermoregulatory centre has been the demonstration that a large proportion (about 40%) of its cells are sensitive to temperature. However, many of these same neurones are also sensitive to osmotic pressure, and to concentration of glucose and steroids. Thus, some of the neural signals that reach the hypothalamus are thermal but many are non-thermal. This supports an interactive arrangement of all the central structures that receive thermal input.

The central neurochemical mechanisms are still not fully understood. In the past a lot of research was carried out to study the role of central amines and prostaglandins in the control of temperature. However, more recently the focus of research has been directed towards the role of central neuropeptides such as adrenocorticotrophic hormone (ACTH) which seems to have some thermoregulatory function.

Non-thermal stimuli may affect thermoregulation since the control of body temperature is brought about, in part, by activating physiological systems which are not specific to temperature control. For example, skeletal muscles of the locomotor system are used for shivering; the vasculature of the cardiovascular system is used to regulate blood flow (and therefore heat flow); the respiratory system is used to mediate evaporative heat loss through panting in some species. These are features of evolutionary development. Thus man does not have a specific thermoregulatory system. Since several CNS sites are involved in thermoregulation, there are no clearly defined efferent motor tracts specific for temperature control. This explains why non-thermal stimuli such as a sudden loud noise cause the thermal sympathetic responses of vasoconstriction and sweating. Indeed, Nelson *et al.* (1984) have suggested that the concept of central thermoregulation is redundant in terms of structure, due to (a) the numerous areas in the central nervous system that respond to temperature, and (b) the multitude of functions performed by the same cells.

Although the hypothalamus has been implicated as the key heat controlling centre in the brain, there are some species, such as the dog and antelope, in which hypothalamic temperature is actually decreased compared with that in the resting state, at times when heat loss mechanisms are operating (increased peripheral blood flow and panting). For example, in the dog, brain temperature decreases during the first few minutes of exercise (figure 6.2). Although rectal temperature increases, this has already been shown to be a poor index of brain temperature. Even during sustained running, hypothalamic temperature is kept well below carotid artery temperature due to the large surface area for evaporative cooling provided by the mouth and by the well developed mechanism for countercurrent heat exchange between the carotid artery and the venous effluent in this species. It has been postulated that a similar mechanism may operate in man in view of the copious sweating seen on the face during exercise and resting thermal stress. There is, however, no conclusive evidence for this.

Owing to the complexity of the central structures involved in temperature control, the peripheral mechanisms, which are more

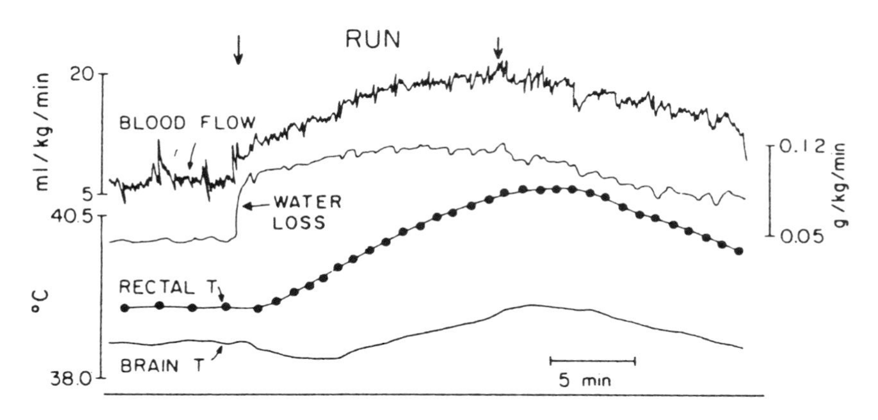

Figure 6.2 Carotid blood flow evaporative heat loss and rectal and hypothalamic temperatures in a running dog. Reproduced with permission from the *Annual Review of Physiology*, Vol. 44 © 1982 by Annual Reviews Inc.

accessible, have been studied more extensively. Most of these peripheral mechanisms are under the control of the autonomic nervous system.

6.5 Autonomic nervous control of thermoregulatory mechanisms and exercise

The rise in core temperature during exercise in cool conditions reaches a plateau if the work load is both constant and lasts for at least 45 min (figure 6.1). This rise in temperature is proportional to the relative work load of an individual (a given percentage of maximum oxygen uptake). It is not proportional to the absolute metabolic rate, except in a single individual, as heat loss mechanisms (in particular sweating) improve with the increase in maximum oxygen uptake, so the relative thermal load is the same. Such a plateau of core temperature is unaffected by a wide range of environmental temperatures, but in hot environments, especially where the humidity is high, the heat loss mechanisms ultimately fail to keep pace with the need to dissipate heat, and core temperature continues to rise. Thus, work in the heat is limited. When these tolerance limits are approached, hyperthermia (a core temperature above 41°C) will occur unless work is stopped voluntarily. The consequences of hyperthermia could be a permanent impairment of thermoregulatory function, although it is not clear whether individuals who suffer heat illness have a pre-existing vulnerability to heat.

6.5.1 BLOOD FLOW

The cutaneous blood vessels are innervated by sympathetic noradrenergic neurones which mediate vasoconstriction (see Chapter 5). In addition, man has an active vasodilator mechanism in skin (as yet unidentified) which is crucial during body heating. This is related to sweating since active vasodilatation is absent in individuals with no sweat glands. Likely candidates for this mechanism are **bradykinin** or **VIP (vasoactive intestinal polypeptide)**, although the evidence is far from conclusive and the mechanism therefore remains obscure.

Blood flow to the skin during exercise is probably related to the percentage of maximum oxygen uptake in view of the latter's relation to core temperature. During short-term vigorous work (up to about

10 min) skin blood flow is reduced, probably due to noradrenaline-mediated vasoconstriction. Since the rise in core temperature is related to relative work load, it follows that it is also related to sympathetic nervous activity. Since sympathetic nervous activity reduces skin blood flow, the ensuing reduction in heat loss could explain why core temperature is related to the percentage of maximum oxygen consumption.

However, during more prolonged, and therefore less vigorous, exercise, skin blood flow is increased. This increase in skin blood flow facilitates heat loss but may lead to a decrease in total peripheral resistance and therefore venous return, cardiac filling pressure and subsequently stroke volume.

During intermittent work of sufficient intensity and duration to raise the core temperature, the skin vasculature dilates during exercise and constricts during rest. The periods of vasoconstriction at rest are thought to have the effect of causing heat storage. This could explain why the core temperature is higher during intermittent exercise than one would predict from steady-state values.

Exercise in the heat

During passive body heating, vasoconstriction in the muscle, splanchnic and renal vascular beds allows an increased proportion of cardiac output to be distributed to the skin. This, together with cutaneous vasodilatation, allows skin blood flow to reach very high levels, with values as high as 3.5 litres m^{-2} min^{-1} being reached during extreme heating. Clearly this requires an increase in cardiac output.

During exercise in the heat it is highly unlikely that levels as high as this are realized due to the demands of the working muscle for blood. During maximal exercise in the heat (actually a lower intensity than in a cool environment), the high demands of working muscle for blood and the continuing demands of a constant blood flow to the brain mean that any further increase in skin blood flow must be diverted away from the viscera and inactive muscle cells and eventually from active muscle. However, the amount of blood available from these tissues is negligible compared with the thermoregulatory demands. Moreover, maximum cardiac output is actually reduced during exercise in the heat as a consequence of a decreased venous return due to reduced venous pressure associated with an increased cutaneous blood flow.

6.5.2 SWEATING

The eccrine sweat glands are innervated by sympathetic cholinergic neurones. Thus, they respond to muscarinic agonists and can be inhibited by anticholinergic drugs such as atropine. There is little evidence of noradrenergic innervation of the sweat glands yet the glands do respond to intradermal injections of adrenaline. The vasoconstriction that accompanies catecholamine-induced sweating would suggest that this mechanism has little thermoregulatory significance.

During exercise, sweat production is related to the absolute work load rather than to the percentage of maximum oxygen consumption. Many of the mechanisms involved in thermoregulation during body heating at rest may be common to those involved in thermoregulation during body heating due to exercise. However, since neurones with thermoregulatory function respond to nonthermal stimuli, some of the thermal responses to exercise may be initiated by mechanisms which do not operate during body heating at rest. This seems to be highly likely for the control of sweating. Sweating during exercise is more profuse than at rest for a similar core temperature. The stimuli for exercise-induced sweating include non-thermal as well as thermal factors arising either from the muscles or from their venous effluent. Evidence for the involvement of a local mechanism in the control of sweating during exercise comes from the observation that sweating can be increased during limb heating even when the warm blood is prevented from reaching the brain by applying an occlusion cuff on an arm. This response is neurogenic since during the occlusion and heating manoeuvre sweating is increased on the trunk and contralateral arm as well as on the ipsilateral arm. Furthermore, sweat production can be switched on and off with the onset and cessation of exercise during intermittent work. This is temporally associated with increases and decreases in femoral venous blood temperature. Such observations are consistent with thermal stimuli in the working muscles, or receptors sensitive to venous blood temperature being involved in the sweating response to exercise.

6.5.3 SHIVERING

Shivering is controlled by somatic nerves but unlike coordinated movement it is largely involuntary. Shivering is a series of muscle contractions which occur in an unsynchronized manner. It is initiated at the spinal level as well as by neurones in the preoptic area of the anterior hypothalamus. As the muscular contractions of shivering are not used to perform useful work, most of the energy is in the form of heat. Shivering, however, is not a very effective means of maintaining core temperature.

During exercise in the cold, shivering may not be necessary due to the elevated metabolism associated with exercise. However, when the exercise intensity is low or reduced by fatigue, shivering will occur. In extreme conditions such as immersion in cold water or during prolonged exposure out of doors in the cold, particularly if wet clothing is worn next to the skin, even shivering may be insufficient to prevent accidental hypothermia (a core temperature less than 35°C).

6.6 Models of thermoregulation

Models of thermoregulation have been developed in an attempt to explain the complexity of the control of body temperature. Inevitably this has been a difficult task and one which is still far from being complete.

6.6.1 SET POINT THEORY

The **set point theory** is that there is a physiological structure or set of structures, including the hypothalamus, which acts as a mammalian thermostat. If the temperature there increases, heat loss is facilitated, and a decrease in hypothalamic temperature initiates heat production and a concomitant reduction in heat loss. Benzinger's now classical theory has been accepted by many people over the last 20 years (see Benzinger, 1969). He argues that the cells of the posterior part of the hypothalamus are inhibited through neural connections with the anterior part of the hypothalamus. Cooling the anterior cells removes this inhibition. Thus, when cold blood reaches the hypothalamus, the posterior cells, no longer inhibited, initiate a cold response (shivering, vasoconstriction and, in neonates and mammals, brown adipose tissue metabolism). Conversely, when the hypothalamus is warmed, the hypothalamic cells initiate a warm response (sweating or panting and vasodilation). In the case of the warm response a clearly defined hypothalamic temperature can be shown for the initiation of sweating. This set point of core temperature seems to be located in the hypothalamus since it can be demonstrated using thermistors located only 2.5 cm from the hypothalamus but not using rectal thermistors. To support this theory Benzinger showed that ingestion of ice caused a decrease in the temperature of blood going to the hypothalamus and a depression of sweating in a hot environment of 45°C.

In resting man the core temperature is some ten times relatively more important that the skin temperature in controlling sweating and twenty times more powerful than the skin temperature in controlling cutaneous blood flow in the heat.

Thus the set point theory involves a negative feedback loop (figure 6.3). According to some proponents of this theory, fever or exercise represent examples of resetting the thermostat to a newly defined higher level.

In contrast to the responses to heating, the set point theory fails to explain adequately the response to cold. Benzinger showed that the responses to cold could be initiated in the absence of a fall in core temperature. Cold water immersion leads to an increase in oxygen consumption due to shivering and vasoconstriction while core temperature is maintained. Thus, in a cold environment the temperature-sensitive receptors in the skin are very powerful in terms of regulating body temperature.

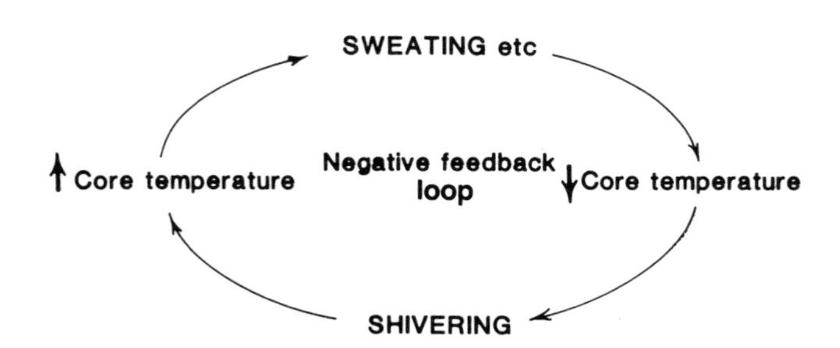

Figure 6.3 Overall scheme of set point theory of thermoregulation.

6.6.2 VARIABLE SET POINT THEORY

The **variable set point theory** holds that the hypothalamus acts as a proportional controller. According to this theory there is a greater interplay between peripheral and central mechanisms than allowed for in the set point theory. Thus, if ambient temperature decreases, the increase in heat production would occur sooner than predicted by the core temperature alone. Similarly, if skin temperature is increased in a warm environment, sweating and cutaneous vasodilatation occur sooner than could be predicted by the core temperature alone (table 6.1 and figure 6.4a).

Table 6.1 Variable set point of core temperature for sweating and shivering

	Sweating response		Shivering response	
Core temperature (°C)	36.8	37.5	37.1	36.5
Skin temperature (°C)	33.0	29.0	20.0	30.0

Hypothetical data from Stainer *et al.* (1984).

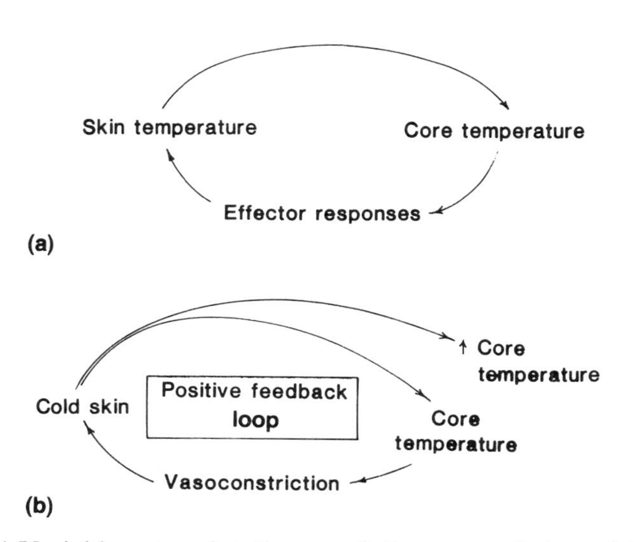

Figure 6.4 Variable set point theory of thermoregulation. (a) Overall scheme of variable set point theory of thermoregulation; (b) positive feedback due to skin temperature.

However, there are two major difficulties with any theory involving set point. One is that such a theory would require swings in temperature, but such variability is not seen. The core temperature exhibits a high degree of stability. There is no evidence of runaway increases and decreases in core or skin temperature while the thermoregulatory mechanisms are being recruited. The second problem is that if skin temperature were as crucial as the variable set point theory suggests, an unbreakable cycle of positive feedback would ensue (assuming there was no decrease in heat production) (figure 6.4b). For these reasons, Gordon and Heath (1983) have proposed an **oscillation theory** of temperature control.

6.6.3 OSCILLATION THEORY

The above set point theories view thermoregulation in terms of the physiological responses to thermal stress. A different view is to regard the thermoregulatory responses in terms of neural activity. This view is central to the oscillation theory of thermoregulation.

All the thermoregulatory mechanisms (metabolic rate, skin temperature, skin blood flow, sweating and shivering) display individually characteristic rates of oscillations which are thought in most cases to originate in the central nervous system. Shivering rhythms may be an exception and may be capable of being initiated in the spinal cord. These oscillations have varied frequencies, with some as low as 0.0002–0.001 Hz for the skin temperature of a finger held in an ice bath and others as high as 18 Hz for cat hind-limb shivering. It is not known how these oscillations might be integrated to bring about thermoregulatory control. However, since some of the fastest rhythms are generated by thermal receptors which have no presynaptic input, the latter are thought to be pacemaker cells, with slower rhythmical activity being generated in the preoptic area of the anterior hypothalamus.

The rhythms in sweating and blood flow are due to rhythmical bursts of firing in the sudomotor and vasomotor postganglionic C fibres, an observation which supports the control of thermoregulation by rhythm-generating mechanisms.

References and further reading

Åstrand, P-O. (1960) Aerobic work capacity in men and women with special reference to age. *Acta Physiol. Scand.*, **49**, Suppl. 169.

Baker, M.A. (1982) Brain cooling in endotherms in heat and exercise. *Ann. Rev. Physiol.*, **44**, 85–96.

Beaumont van, W. and Bullard, R.W. (1966) Sweating: its rapid response to muscular work. *Science*, **141**, 643–646.

Beaumont van, W. and Bullard, R.W. (1966) Sweating exercise stimulation during circulatory arrest. *Science*, **152**, 1521–1523.

Benzinger, T.H. (1969) Heat regulation: homeostasis of central temperature in man. *Physiol. Rev.*, **49**, 671–759.

Bini, G., Hagbarth, K-E., Hynninen, P. and Wallin, B.G. (1980a) Thermoregulatory and rhythm-generating mechanisms governing the sudomotor and vasoconstrictor outflow in human cutaneous nerves. *J. Physiol.*, **306**, 537–552.

Bini, G., Hagbarth, K-E., Hynninen, P. and Wallin, B.G. (1980b) Regional similarities and differences in thermoregulatory vaso- and sudomotor tone. *J. Physiol.*, **306**, 553–565.

Boulant, J.A. and Dean, J.B. (1986) Temperature receptors in the central nervous system. *Ann. Rev. Physiol.*, **48**, 639–654.

Brengelmann, G.L., Freund, P.L., Rowell, L.B., Oleurd, J.E. and Kraning, K.K. (1981) Absence of active cutaneous vasodilation associated with congenital absence of sweat glands in man. *Am. J. Physiol.*, **240**, H571–H575.

Burton, A.C. (1935) Human calorimetry II: the average temperature of the body. *J. Nutr.*, **9**, 261–280.

Burton, A.C. (1939) Range and variability of blood flow in human fingers and the vasomotor regulation of body temperature. *Am. J. Physiol.*, **127**, 437–453.

Buskirk, E.R. (1977) Temperature regulation with exercise. *Ex. Sports Sci. Rev.*, **5**, 45–88.

Clark, R.P., Mullan, B.J. and Pugh, L.G.C.E. (1977) Skin temperature during running — a study using infra-red colour thermography. *J. Physiol.*, **267**, 53–62.

Cooper, K.E., Cranston, W.I. and Snell, E.S. (1964) Temperature in the external auditory meatus as an index of central temperature changes. *J. Appl. Physiol.*, **19**, 1032–1035.

Cranston, W.I., Gerbrandy, I. and Snell, E.S. (1954) Oral, rectal and oesophageal temperatures and some factors affecting them in man. *J. Physiol.*, **126**, 347–358.

DuBois, D. and DuBois, E.F. (1916) Clinical calorimetry. X. A formula to estimate the approximate surface area if height and weight be known. *Arch. Intern. Med.*, **17**, 863.

Edwards, R.J., Belyavin, A.J. and Harrison, M.H. (1978) Core temperature measurement in man. *Aviat. Space Environ. Med.*, **49**, 1289–1294.

Fox, R.H., Solman, A.J., Isaacs, R., Fry, A.J. and Macdonald, I.C. (1973)

A new method for monitoring deep body temperature from the skin surface. *Clin. Sci.*, **44**, 81–86.

Gisolfi, C. and Robinson, S. (1970) Central and peripheral stimuli regulating sweating during intermittent work in men. *J. Appl. Physiol.*, **29**, 761–768.

Gordon, C.J. and Heath, J.E. (1980) Slow bursting thermal sensitive neurons in the preoptic area of the rabbit. *Brain Res. Bull.*, **5**, 515–518.

Gordon, C.J. and Heath, J.E. (1983) Reassessment of the neural control of body temperature: importance of oscillating neural and motor components. *Comp. Biochem. Physiol.*, **74A**, 479–489.

Gordon, C.J. and Heath, J.E. (1986) Integration and central processing in temperature regulation. *Ann. Rev. Physiol.*, **48**, 595–612.

Harada, E. (1971) A characteristic pattern of fluctuation in the skin temperature of the rabbit's ear in response to alteration of the environmental temperature. *J. Physiol. Soc. Jap.*, **33**, 303–316.

Hardy, J.D. and DuBois, E.F. (1938) Basal metabolism, radiation, convection and vaporization at temperatures of 22–35°C. *J. Nutr.*, **15**, 477–497.

Hardy, J.D., Milharat, A.T. and DuBois, E.F. (1941) Basal metabolism and heat loss of young women at temperatures from 22 to 25°C. *J. Nutr.*, **21**, 383–404.

Hurley, H.J. and Witowski, J.A. (1961) Mechanism of epinephrine-induced eccrine sweating in human skin. *J. Appl. Physiol.*, **16**, 652–654.

Lipton, J.M. and Clark, W.G. (1986) Neurotransmitters in temperature control. *Ann. Rev. Physiol.*, **48**, 613–623.

Mead, J. and Bonmarito, C.L. (1949) Reliability of rectal temperature as an index of internal body temperature. *J. Appl. Physiol.*, **2**, 97–109.

Nelson, D.O., Heath, J.E. and Prosser, C.L. (1984) Evolution of temperature regulating mechanisms. *Am. Zool.*, **24**, 791–807.

Rawson, R.O. and Hammel, H.T. (1963) Hypothalamic and tympanic membrane temperatures in Rhesus monkey. *Fed. Proc.*, **22**, 283.

Robinson, S., Meyer, F.R., Newton, J.L., Ts'ao, C.H. and Holgersen, L.O. (1965) Relations between sweating cutaneous blood flow, and body temperature in work. *J. Appl. Physiol.*, **20**, 575–582.

Rowell, L.B. (1974) Human cardiovascular adjustments to exercise and thermal stress. *Physiol. Rev.*, **54**, 75–159.

Rowell, L.B. (1983) Cardiovascular aspects of human thermoregulation. *Circ. Res.*, **52**, 367–379.

Saltin, B. and Hermansen, L. (1966) Esophageal, rectal and muscle temperature during exercise. *J. Appl. Physiol.*, **21**, 1757–1762.

Shapiro, Y., Magazanik, A., Udassin, R., Ben-Baruch, G., Shvartz, E. and Shoenfeld, Y. (1979) Heat intolerance in former heatstroke patients. *J. Int. Med.*, **99**, 913–916.

Stainer, M.W., Mount, L.E. and Bligh, J. (1984) *Energy Balance and Temperature Regulation*, Cambridge University Press, Cambridge.
Stuart, D.G., Eldred, E., Hemingway, A. and Kawamura, Y. (1966) The rhythm of shivering. I. General sensory contributions. *Am. J. Phys. Med.*, **45**, 61–74.
Teichner, W.H. (1958) Assessment of mean body surface temperature. *J. Appl. Physiol.*, **12**, 169–176.
Warndorff, J.A. and Neefs, J. (1971) A quantitative measure of sweat production after local injection of adrenaline. *J. Invest. Derm.*, **56**, 384–386.
Wyndham, C.H. (1965) Role of skin and core temperature in man's temperature regulation. *J. Appl. Physiol.*, **20**, 31–36.

Chapter 7

Factors affecting autonomic nervous activity

In this chapter the effects of ageing, sex, training and drugs on autonomic nervous activity during exercise are considered.

7.1 Ageing

7.1.1 SYMPATHOADRENAL ACTIVITY

Circulating catecholamine concentrations in individuals at rest increase with age, and by the age of 70 years they are some 50–100% greater than at the age of 20 years. This could be due to an increase in sympathoadrenal activity, or to a reduction in plasma catecholamine removal or both (see Chapter 2). Measurements of noradrenaline kinetics using infusions of tritiated noradrenaline suggest that the age-related rise in plasma catecholamines is primarily the result of reduced clearance rather than an increase in production, although the evidence is inconclusive. At least one study has shown only an elevated plasma adrenaline while failing to detect increased plasma noradrenaline concentrations with age at rest. The variable findings for noradrenaline are consistent with the high degree of variability in directly measured sympathetic nervous activity between subjects despite a trend for an increase with ageing.

Older individuals, aged 66–77 years, display an augmented adrenaline and noradrenaline response to graded dynamic exercise up to maximum work capacity compared with young subjects aged 22–37 years (Fleg *et al.*, 1985). Whilst this may be expected at the same absolute work loads in view of the well established decline in work capacity with age, it persists even when the catecholamine data are related to relative work load. At maximum exercise capacity the difference between groups for both amines was significant at the 5% level in the latter study. This contrasts with the data of Lehmann and Keul (1986) who showed significantly higher resting noradrenaline and adrenaline levels in the older (66 ± 6 years) compared with younger (25 ± 3 years) group but no significant differences in either amine at maximal exercise. Interestingly, the absolute values for plasma catecholamine concentrations reported by Lehmann and Keul are very much higher than those reported by Fleg *et al.*. Indeed the mean levels in maximal exercise are high enough to be considered pathologically high. The differences between the two studies are unlikely to be due to the mode of exercise as Lehmann and Keul's subjects used a bicycle ergometer whereas those of Fleg *et al.* used a treadmill. Since treadmill exercise is known to elicit a higher maximum oxygen consumption than bicycle work, one would have expected Fleg *et al.* rather than Lehmann and Keul, to have reported the higher values. Alternatively (and most likely) the differences could reflect interlaboratory discrepancies in the measurement of catecholamines (see Chapter 2), despite the fact that both used the same (radioenzymatic) methods. Another observation of these data that seems important is that, if the absolute values in each study are reliable, then Lehmann and Keul's young subjects have higher mean circulating catecholamines both at rest and during exercise than the older group of Fleg *et al.* (figure 7.1).

It appears that the higher circulating catecholamines with ageing do not lead to augmented sympathetic responses to exercise. Rather, with advancing age there is less reliance on a catecholamine-mediated increase in heart rate and reduction in end-systolic volume and a greater reliance on the Frank–Starling mechanism for increasing stroke volume during exercise. This is consistent with a decreased sensitivity of the β_1-adrenoceptors or a decrease in the supply of the second messenger, cAMP. The latter is thought to contribute to the decrease in β_1-adrenoceptor-mediated lipolysis seen in isolated adipocytes from old compared with young rats.

7.1.2 SYMPATHETIC CHOLINERGIC ACTIVITY

Ageing is associated with a smaller sweating response to both thermal (heat) and chemical (acetylcholine) stimuli. During exercise it is important that the sweating response to exercise is related to relative

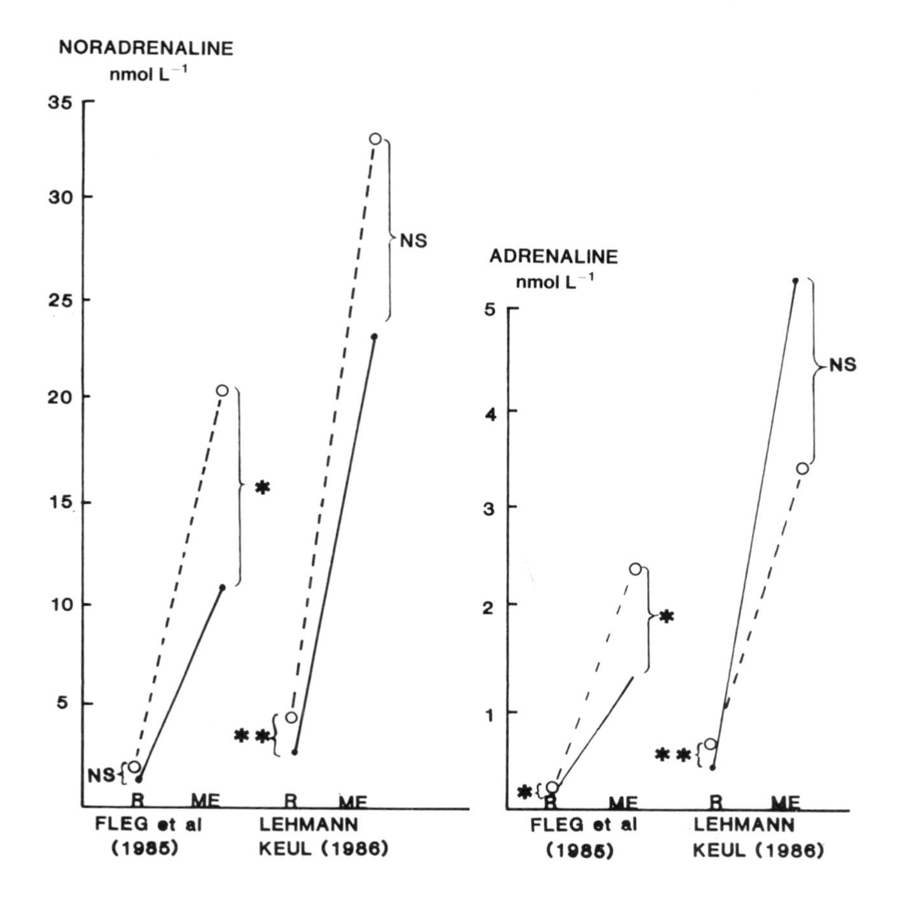

Figure 7.1 Mean circulating catecholamines at rest and during maximal exercise in young (closed circles) and old (open circles) subjects. R = Rest; ME = maximal exercise; * = $P < 0.05$, ** = $P < 0.01$, NS = no significant difference.

work load since at absolute work loads the older (less fit) individual is likely to appear to have a reduced sweating response. Such a finding would be due to a decrease in work capacity rather than a decline in sweat gland function. Nevertheless there are conflicting reports in the literature as to whether sweat production rate is reduced with age during steady-state exercise. With intermittent exercise, forearm sweat production rate is lower in middle-aged compared with young women, and this may be related to changes in central thermoregulatory function with age.

7.2 Sex

7.2.1 SYMPATHOADRENAL ACTIVITY

Most human studies have been carried out on men. However, the few studies that have examined the possibility of sex differences in the catecholamine response to exercise suggest that at the same relative work load the catecholamine levels are similar between the sexes in both isotonic and isometric exercise. However, interestingly in isometric exercise involving 30% of maximal handgrip until exhaustion, men show a higher adrenaline response in the first minute of exercise compared with women. Moreover, if changes rather than absolute levels of catecholamines are examined, men have a greater increase in all three catecholamines (noradrenaline, adrenaline and dopamine) than women at every minute of 5 min of exercise. This would be consistent with either an augmented sympathetic response in men or reduced noradrenaline clearance. Despite this the heart rates are the same; this could be coincidental, or the baroreflex could be responsible for modulating the vagal withdrawal, such that a β-agonism in the heart would give rise to less baroreflex activation and so less of a change in vagal tone. This could be a result of sex differences in the central control mechanisms. Another possibility is that adrenal medullary catecholamine production is lower in women as a result of an oestrogen-mediated inhibition of the cholinergic stimulation of the adrenal medulla.

7.2.2 SYMPATHETIC CHOLINERGIC ACTIVITY

Since sweat production may be an index of sympathetic cholinergic activity, the well established earlier sweating response and greater sweating capacity of men compared with women could be interpreted as an augmented sympathetic cholinergic response. The earlier sweating response to exercise may be necessary in men who do not have the thermoregulatory advantage of the oestrogen-mediated increase in cutaneous vascularization. However, the increased sweating capacity is probably a function of the increased work capacity of men rather than any thermoregulatory differences between the sexes. At relative work loads there is no difference in the sweating or core temperature responses of men and women.

Nevertheless there are differences in sweat-gland density and sweat production between the sexes, with women having a greater density of sweat glands than men but men having a higher production of sweat per gland than women. The density of sweat glands seems to be inversely related to body surface area, and the smaller surface area of women compared with men could explain the sex differences in sweat-gland density.

7.3 Training

7.3.1 SYMPATHOADRENAL ACTIVITY

At any given absolute submaximal work load there is a smaller plasma catecholamine response in trained compared with untrained individuals, whereas there is no difference between two such groups when the catecholamine response is assessed in relation to relative work loads. The absolute differences in plasma catecholamines between trained and untrained individuals persist (albeit less markedly) even when trained athletes exercise with untrained muscles. This suggests that the sympathetic drive is not greatly influenced by the local muscular adaptations to training, but central nervous factors affecting medullary neurones may possibly be contributors.

The decrease in plasma catecholamines at a given absolute work load occurs quickly, with maximal reductions being recorded after only three weeks of vigorous training. This decrease in plasma catecholamines could explain some of the decrease in heart rate at a given absolute work load with training. However, since the exercising heart rate takes five weeks of endurance training to reach minimal levels, other factors must contribute (figure 7.2). Such factors could include increased parasympathetic activity as well as a decrease in cardiac β-adrenoceptor density or affinity for catecholamines.

To evaluate the effects of training on adrenaline-mediated physiological responses, Svedenhag *et al.* (1986) gave adrenaline infusions to well-trained runners and to sedentary individuals before and after a period of training. The infusion of 0.1–0.6 nmol

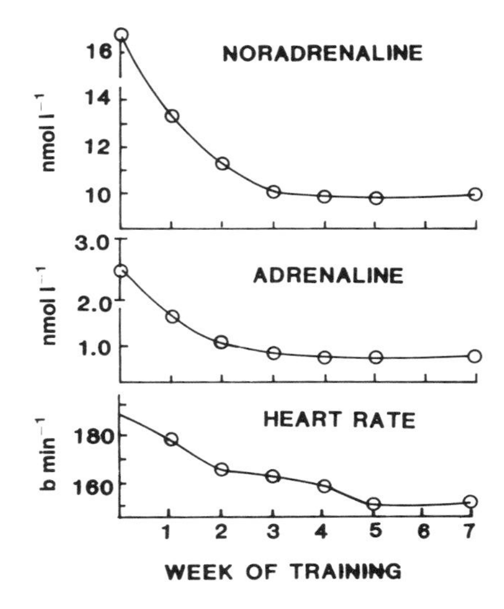

Figure 7.2 Effect of training on plasma catecholamines and heart rate during exercise. Figure adapted from Winder *et al.* (1978).

adrenaline per kilogram per minute produced lower circulating adrenaline in the well-trained compared with sedentary subjects. However, after training, the sedentary subjects showed no such attenuation of their plasma adrenaline concentration. The reasons for these results are not clear. One possibility is that the well-trained runners had a higher blood volume (thus effectively diluting the catecholamine concentration in blood), and the second is that training lasting several years (but not weeks) enhances adrenaline clearance. However, at rest there is no difference between trained and untrained individuals in terms of adrenaline clearance rate.

The limitations of using plasma catecholamines to assess sympathetic nervous activity have already been outlined in Chapter 2. Such limitations could explain the apparently contradictory results of direct recordings of sympathetic nerve activity to skeletal muscle (peroneal nerve). Such studies reveal no significant changes in muscle nerve activity with training. However, in the latter study there was no significant reduction in resting heart rate with training (consistent with no significant change in cardiac sympathetic nervous activity). This might suggest that the exercise regime was inadequate to produce significant training effects. Nevertheless, there were similarly no differences in muscle sympathetic activity between well-trained cyclists and sedentary individuals. The latter studies were carried out in resting subjects and it may be that any differences in sympathetic nervous activity between trained and untrained individuals only become manifest during physiological stress or during physical activity. Evidence for the former includes the observation that trained individuals display an enhanced adrenaline response (but not an enhanced noradrenaline response) to hyper-glucagonaemia, hypercapnia and hypoxia compared with sedentary individuals despite similar values of plasma glucagon, pO_2 and pCO_2 between the two groups. This is attributed to augmented secretion of adrenaline rather than a decreased clearance rate since the latter is unchanged at rest by training. Training also reduces the sensitivity of the pancreatic β cells so that circulating insulin levels are reduced in trained individuals in response to glucagon infusion. This reduction in circulating insulin is compensated for by an increased sensitivity of the target tissues to insulin.

7.3.2 SYMPATHETIC CHOLINERGIC ACTIVITY

Training leads to an improvement in sweat-gland activity, and this is due to peripheral mechanisms, namely an increase in cholinergic sensitivity, as well as an enlargement of the sweat glands. There is a linear relationship between maximum sweat production induced by pilocarpine (an acetylcholine agonist), iontophoresis and maximum oxygen uptake such that, for every increase in maximum oxygen uptake of 10 ml kg^{-1} min^{-1} above 30 ml kg^{-1} min^{-1}, maximum sweat rate increases by 1.6 g m^{-2} min^{-1}. This means that the sedentary individual with a low maximum oxygen uptake of 30 ml kg^{-1} min^{-1} has a predicted maximum sweat rate of 1.6 g m^{-2} min^{-1}, and this maximum rate of sweating is 500% less than the maximum sweat rate of 8.0 g m^{-2} min^{-1} predicted for a well-trained marathon runner with a maximum oxygen consumption of 70 ml kg^{-1} min^{-1}

7.4 Drugs

Drugs have played a vital role in many experiments to further understanding of the autonomic nervous system. Pharmacological compounds have been used to selectively block or stimulate ganglionic or postganglionic neurotransmission in sympathetic or parasympathetic neurones, thereby elucidating the roles of different components of the autonomic nervous system. A major limitation in the use of drugs in this latter regard is that their actions are not necessarily organ specific. While a drug may have a primary effect on a given organ, there may be side-effects on other organs. Many of these drugs are used in clinical practice for the treatment of conditions in which autonomic activity is implicated. Drugs affecting autonomic function are also used for pleasure (for example nicotine in cigarettes) or by élite athletes in an attempt to improve physical performance (typically the amphetamines).

It is beyond the scope of this book to examine comprehensively the effect of drugs on the autonomic nervous system. The interested reader should consult a pharmacology text, such as Crossland's *Lewis's Pharmacology*, for a detailed account of this topic. Some consideration is given below to studies of β-adrenoceptor antagonists during exercise and training in view of the considerable interest that has been shown in these compounds in recent years.

7.4.1 EFFECTS OF β-ADRENOCEPTOR ANTAGONISM DURING EXERCISE

It is well known that administration of β-adrenoceptor antagonists reduces endurance exercise capacity but that these drugs do not seem to affect neuromuscular skill or activities which predominantly require muscle strength. The limitation to endurance exercise during β-adrenoceptor antagonism includes a decrease in muscle blood flow. This is not due to a direct effect on muscle perfusion since the β_2-adrenoceptor mechanism for increasing blood flow is not involved in the control of blood flow during exercise (Chapter 5). Any decrease in muscle blood flow is secondary to a decrease in cardiac output as a result of a decrease in the chronotropic and inotropic activity of the heart. Despite the decrease in blood flow, oxygen delivery to the muscle cells is not impaired as oxygen extraction increases. At high levels of exercise, when the maximal $a - \bar{v}\ O_2$ value has been achieved, exercise capacity is limited by the reduced cardiac output. At submaximal work loads, exercise may be limited by antagonism of β_1-adrenoceptor-mediated lipolysis leading to increased reliance on carbohydrate metabolism.

7.4.2 EFFECTS OF β-ADRENOCEPTOR ANTAGONISM ON THE PHYSIOLOGICAL RESPONSES TO TRAINING

Despite the significant role of the sympathoadrenal system in periods of acute exercise, long-term β-blockade does not seem to affect the improvement in maximum oxygen consumption with training (Svedenhag *et al.*, 1984a). The latter study involved 8 weeks of training by healthy subjects who were either taking a placebo or approximately 2 mg propranolol per kilogram per day. This dose is comparable to moderate β-blockade typical of treatment for hypertension. The subjects underwent their post-training physiological measurements 4 days after they had stopped taking propranolol. In this way any effect of the drug on training independently of its known effects on acute exercise could be assessed. At submaximal work loads there was a smaller increase in heart rate in the placebo and propranolol groups. At higher work loads this depression of heart rate was more marked in the control, compared with the propranolol group. This may be related to the greater contribution of sympathetic drive, compared with reduction in vagal tone, at higher work loads in the unblocked state. This would be consistent with the observation in rats that training bradycardia is due mainly to a reduction in sympathetic tone, rather than to vagal excitation.

The results of the study by Svedenhag *et al.* in men contrast with results of a series of studies in rats by Harri (1979, 1980) and Harri and Narvola (1979). Harri demonstrated that training for seven weeks while taking propranolol led to a higher resting and submaximal exercising heart rate and less marked metabolic and cardiovascular improvements compared with control rats given a placebo. The discrepancies could be due to differences in propranolol dose (Harri used 10 mg kg^{-1}), differences between species, and the fact that Harri retested the rats only 24 h after the last dose of propranolol.

References and further reading

Bittel, J. and Henane, R. (1975) Comparison of thermal exchanges in men and women under neutral and hot conditions. *J. Physiol.*, **250**, 475–489.

Buono, M.J. and Sjoholm, N.T. (1988) Effect of physical training on peripheral sweat production. *J. Appl. Physiol.*, **65**, 811–814.

Cable, N.T. and Green, J.H. (1989) The influence of bicycle exercise, with or without hand immersion in cold water, on forearm sweating in young and middle-aged women. *J. Physiol.* (abstract). In press.

Crossland, J. (1980) *Lewis's Pharmacology*, 5th edition, Churchill Livingstone, Edinburgh.

Cryer, P.E. (1980) Physiology and pathophysiology of the human sympathoadrenal system. *New Engl. J. Med.*, **303**, 436–444.

Davies, C.T.M. (1979) Thermoregulation during exercise in relation to sex and age. *Eur. J. Appl. Physiol.*, **42**, 71–79.

Day, M.D. (1979) *Autonomic Pharmacology: Experimental and Clinical Aspects*. Churchill Livingstone, Edinburgh.

Esler, M., Skews, H., Leonard, P., Jackman, G., Bobik, A., Korner, P. (1981) Age-dependence of noradrenaline kinetics in normal subjects. *Clin. Sci.*, **60**, 217–219.

Fleg, J.L., Tzankoff, S.P. and Lakatta, E.G. (1985) Age-related augmentation of plasma catecholamines during dynamic exercise in healthy males. *J. Appl. Physiol.*, **59**, 1033–1039.

Galbo, H. (1983) *Hormonal and Metabolic Adaptation to Exercise*, Georg Thieme Verlag, Stuttgart. [See especially section 2.2, Factors influencing sympathoadrenal activity during exercise, pp. 5–21].

Galbo, H., Hedeskov, C.J., Capito, K. and Vinten, J. (1981) The effect of physical training on insulin secretion of rat pancreatic islets. *Acta Physiol. Scand.*, **111**, 75–79.

Galbo, H., Kjær, M. and Secher, N.H. (1987) Cardiovascular, ventilatory and catecholamine responses to maximal dynamic exercise in partially curarized man. *J. Physiol.*, **389**, 557–568.

Gilman, A., Goodman, L.S., Rall, T.W. and Murad, F. (1985) *Goodman and Gilman's The Pharmacological Basis of Therapeutics*, 7th edition, Macmillan Publishing Company, New York.

Greenleaf, J.E., Castle, B.L. and Ruff, W.K. (1972) Maximal oxygen uptake, sweating and tolerance to exercise in the heat. *Int. J. Biometeorol.*, **16**, 375–387.

Harri, M.N.E. (1979) Physical training under the influence of beta blockade in rats: II Effects on vascular reactivity. *Eur. J. Appl. Physiol.*, **42**, 151–157.

Harri, M.N.E. (1980) Physical training under the influence of beta blockade in rats: III Effects on muscular metabolism. *Eur. J. Appl. Physiol.*, **45**, 25–31.

Harri, M.N.E. and Narvola, I. (1979) Physical training under the influence of beta blockade in rats: effect on adrenergic responses. *Eur. J. Appl. Physiol.*, **41**, 199–210.

Hoffman, B.B., Chang, H., Farahbakhsh, Z.T. and Reaven, G.M. (1984) Age-related decrement in hormone-stimulated lipolysis. *Am. J. Physiol.*, **247**, E772–E777.

Juhlin-Dannfelt, A. (1983) β-Adrenoceptor blockade and exercise: effects on endurance and physical training. *Acta Med. Scand.* Suppl., **672**, 49–54.

Kjær, M., Christensen, N.J., Sonne, B., Richter, E.A. and Galbo, H. (1985) Effect of exercise on epinephrine turnover in trained and untrained male subjects. *J. Appl. Physiol.*, **59**, 1061–1067.

Kjær, M. and Galbo, H. (1988) Effect of physical training on the capacity to secrete epinephrine. *J. Appl. Physiol.*, **64**, 11–16.

Lehmann, M. and Keul, J. (1986) Age-associated changes of exercise-induced plasma catecholamine responses. *Eur. J. Appl. Physiol.*, **55**, 302–306.

Lin, Y-C. and Horvath, S.M. (1972) Autonomic nervous control of cardiac frequency in the exercise-trained rat. *J. Appl. Physiol.*, **33**, 796–799.

Mikines, K.J., Dela, F., Sonne, B., Farrell, P.A., Richter, E.A. and Galbo, H. (1987) Insulin action and secretion in man. Effects of different levels of physical activity. *Can. J. Sports Sci.*, **12**, Suppl. 1, 113–116.

Rodeheffer, R.J., Gerstenbilth, G., Becker, L.G., Fleg, J.L., Weisfeldt, M.L. and Lakatta, E.G. (1984) Exercise cardiac output is maintained with advancing age in healthy human subjects: cardiac dilatation and increased stroke volume compensate for a diminished heart rate. *Circulation*, **69**, 203–213.

Rowe, J.W. and Troen, B.R. (1980) Sympathetic nervous system and ageing in man. *Endocrine Rev.*, **1**, 167–179.

Sanchez, J., Pequignot, J.M., Peyrin, L., Monod, H. (1980) Sex differences in the sympatho-adrenal response to isometric exercise. *Eur. J. Appl. Physiol.*, **45**, 147–154.

Sato, K. and Sato, F. (1983) Individual variations in structure and function of human eccrine sweat glands. *Am. J. Physiol.*, **245**, R203–R208.

Sundlöf, G. and Wallin, B.G. (1978) Human muscle nerve sympathetic activity at rest. Relationship to blood pressure and age. *J. Physiol.*, **274**, 383–397.

Svedenhag, J., Henriksson, J., Juhlin-Dannfelt, A. and Asano, K. (1984a) Beta-adrenergic blockade and training in healthy men — effects on central circulation. *Acta Physiol. Scand.*, **120**, 77–86.

Svedenhag, J., Wallin, B.G., Sundlöf, G. and Henriksson, J. (1984b) Skeletal muscle sympathetic activity at rest in trained and untrained subjects. *Acta Physiol. Scand.*, **120**, 499–504.

Svedenhag, J., Martinsson, A., Ekblom, B. and Hjemdahl, P. (1986) Altered cardiovascular responsiveness to adrenaline in endurance-trained subjects. *Acta Physiol. Scand.*, **126**, 539–550.

Trap-Jensen, J., Christensen, N.J., Clausen, J.P., Rasmussen, B. and Klausen, K. (1973) Arterial noradrenaline and circulatory adjustment to strenuous exercise with trained and nontrained muscle groups. *Physical Fitness, Proceedings of the XXVth International Congress of Physiological Sciences*, University of Karlova Press, Prague.

Weichman, B.E. and Borowitz, J.L. (1979) Effects of steroid hormones and diethylstilbestrol on adrenomedullary catecholamine secretion. *Pharmacology*, **18**, 195–201.

Weinman, K.P., Slabochova, Z., Bernauer, E.M., Morimoto, T. and Sargent, F. (1967) Reactions of men and women to repeated exposure to humid heat. *J. Appl. Physiol.*, **22**, 533–538.

Winder, W.W., Hagberg, J.M., Hickson, R.C., Ehsani, A.A. and McLane, J.A. (1978) Time course of sympathoadrenal adaptation to endurance exercise training in man. *J. Appl. Physiol.*, **45**, 370–374.

Ziegler, M.G., Lake, C.R. and Kopin, I.J. (1976) Plasma noradrenaline increases with age. *Nature*, **261**, 333–335.

Chapter 8

Exercise and disease

This chapter will consider diabetes mellitus, obesity, hypertension, atherosclerosis and ischaemic heart disease and the effect of these conditions on autonomic nervous activity and exercise.

8.1 Diabetes mellitus

The incidence of diabetes mellitus is around 2.5% in Europe and the United States of America, with about 0.5% being type I or juvenile onset and 2% being type II or maturity onset. The condition stems from insulin insufficiency associated with hyperglycaemia (random blood glucose 17–65 mmol litre^{-1}), glycosuria, ketosis, acidosis and weight loss despite hyperphagia (increased appetite), and in type I but less commonly in type II diabetes eventually coma. The hyperglycaemia is due to the absence not only of insulin-mediated glucose transport into cells but also of insulin-mediated inhibition of hepatic glucose production. Thus, there is a concomitant decrease in hepatic glycogen content. Despite a plentiful supply of blood glucose there is drastic glucopenia. To meet metabolic demands for energy, protein and fat are used.

Both **proteolysis** (breakdown of proteins to individual amino acids) and the use of amino acids as a precursor for gluconeogenesis are increased in uncontrolled diabetes, and the increase in hepatic glucose production serves to exacerbate the existing hyperglycaemia.

Insulin normally inhibits hormone-sensitive lipase in adipose tissue. In uncontrolled diabetes the absence of insulin leads to an enhanced breakdown of triacylglycerol to free fatty acids (FFA) and glycerol. Thus, blood FFA and glycerol levels are elevated. The FFA are metabolized via β-oxidation (see Chapter 4) but the supply of FFA exceeds the capacity for their oxidation. Excess acetyl coA is converted to acetoacetyl coA to form the ketone bodies, acetoacetate and β-hydroxybutyrate. When acetoacetate is formed in large amounts, the volatile ketone body acetone is also formed (figure 8.1). Ketone bodies are an important source of metabolic energy in diabetes (see figure 8.2), but they also disturb the acid–base balance.

The H^+ ions from acetoacetic acid and β-hydroxybutyrate may accumulate sufficiently to exceed the buffering capacity of the tissues, thereby resulting in metabolic acidosis. The increased H^+ stimulates the peripheral chemoreceptors leading to reflex firing of the respiratory neurones in the medulla (see Chapter 1), causing hyperventilation. This increases respiratory water loss and contributes to dehydration.

Further dehydration occurs due to the increased osmotic concentration of the extracellular fluid which draws water from the

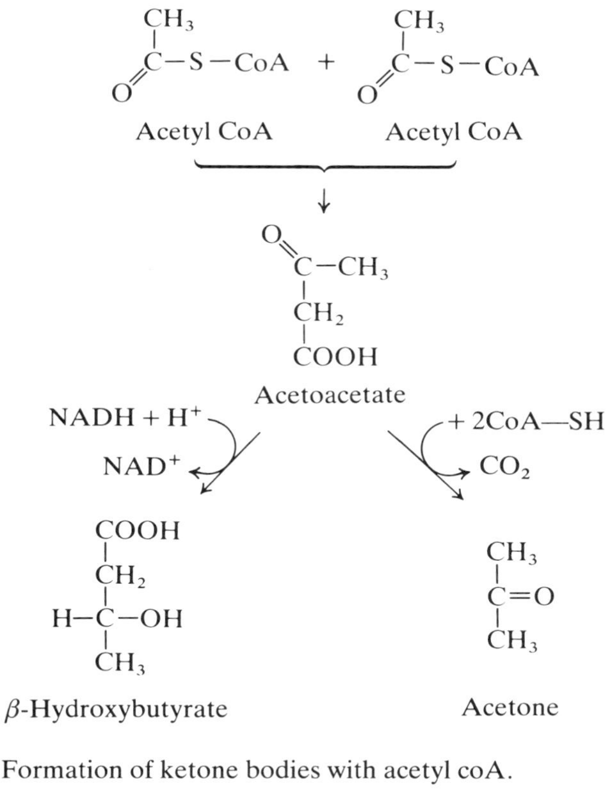

Figure 8.1 Formation of ketone bodies with acetyl coA.

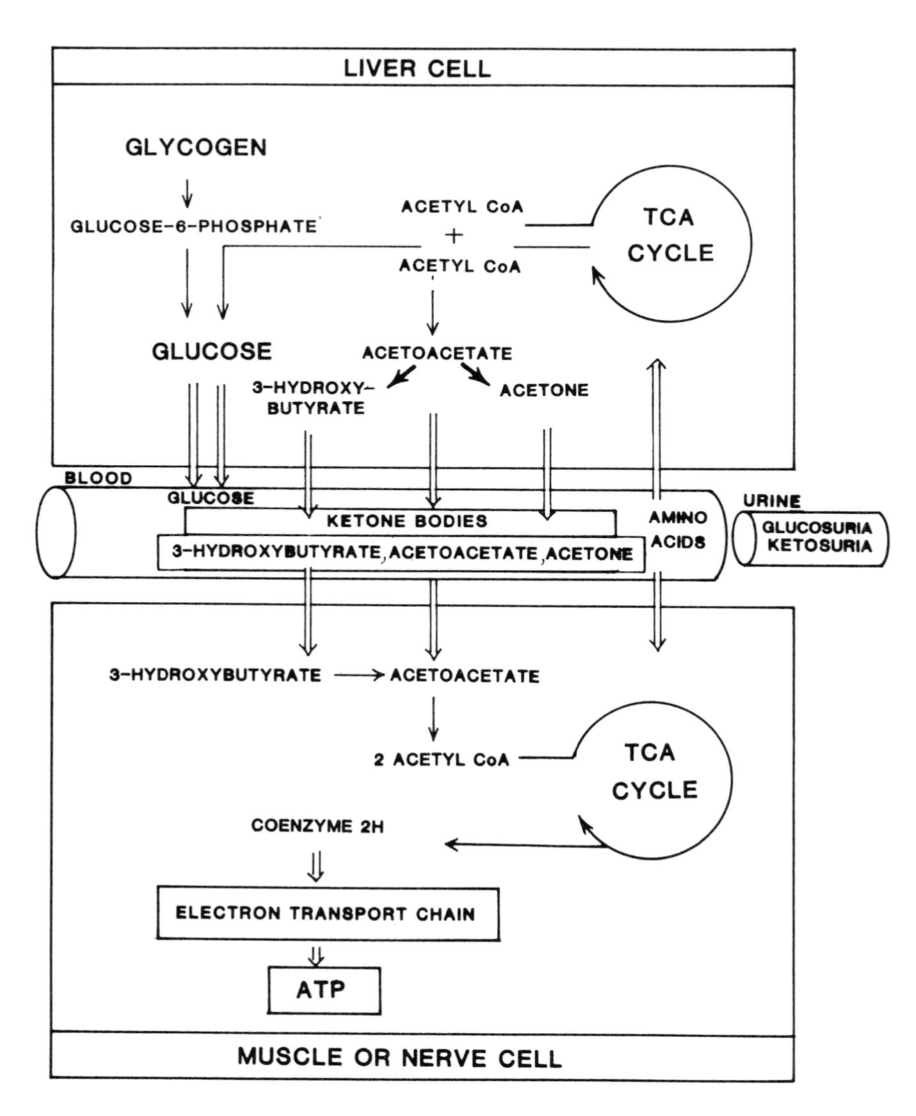

Figure 8.2 Generation of ATP in diabetic ketoacidosis.

cells, reducing intracellular volume. When the blood glucose rises above about 20 mmol litre^{-1}, some glucose appears in the urine (glucosuria). The glucose-induced increase in osmolarity in the tubules prevents fluid reabsorption and thereby produces an increased urine volume (polyuria). This cascade of metabolic events may lead to coma and therefore is a critical condition for the patient.

The cause of diabetes in childhood or adolescence (juvenile-onset diabetes) is unclear with evidence for viral destruction of the pancreatic β cells, genetic disposition and environmental causes. Reduced glucose tolerance is also associated with ageing and is exacerbated by high intake of sugar in the diet. Thus, maturity-onset diabetes is usually traced to obesity, an inadequate pancreatic β cell response and perhaps to high glucose loads in the diet.

Type I diabetes is treated with subcutaneous injections of insulin by the patient which lower blood glucose in around 2 h, depending on the type of insulin. It is important to control diabetes, not only because of the short-term problem of ketoacidosis but also because of the enhanced risk of atherosclerosis which results from the increased reliance on fat metabolism when glucose is unavailable to the cells. The assessment of how much insulin is required is often difficult, and many patients are poorly controlled. Insulin excess, causing hypoglycaemia (blood glucose < 2.0 mmol litre^{-1}), is therefore not uncommon in the management of diabetes mellitus.

8.1.1 INSULIN-INDUCED HYPOGLYCAEMIA

Nerve cells predominantly use carbohydrate. Thus when insulin is given in excess to cause hypoglycaemia, the blood supply of glucose to neural tissue is reduced. Neurones with high metabolic rates are the first to display signs of glucopenia. The first neural signs of hypoglycaemia are therefore the result of a glucose deficit in the cerebral cortex and include giddiness and mental confusion. Euglycaemia is restored by the action of counter-regulatory hormones, principally glucagon and to some extent adrenaline.

Glucagon release is stimulated by a decrease in blood glucose of only 2.2 mmol litre^{-1} from normal values. It acts via the cAMP system to activate hepatic phosphorylase and thereby hepatic glycogenolysis and glucose production. Glucagon is usually the most important counter-regulatory hormone. However, in diabetes the pancreatic α cell response may be impaired so that the plasma catecholamines become crucial for counter-regulation of blood glucose.

Hypoglycaemia is a potent stimulus for increasing plasma catecholamines. Plasma adrenaline increases by around 15 times above baseline and plasma noradrenaline may double. Despite these dramatic rises in circulating catecholamines, they do not normally play the primary role in restoring blood glucose. Evidence for this comes from the use of α- or β-adrenoceptor antagonists which have no effect on blood glucose kinetics after insulin administration. This is consistent with the normal glucose recovery immediately following hypoglycaemia in adrenalectomized patients. However, the delayed glucose response seems to involve catecholamine action. During hypoglycaemia, muscle glycogenolysis is stimulated by a β_2-adrenoceptor mechanism to produce lactate as a precursor for hepatic gluconeogenesis, thereby causing an increase in hepatic glucose output. The administration of propranolol (a non-selective β-adrenoceptor antagonist) is associated with a delayed rise in blood glucose following hypoglycaemia, which is attributed to an inhibition of lactate production.

The increase in plasma catecholamines causes an increase in resting metabolic heat production. Using selective and non-selective β-adrenoceptor antagonists (metoprolol and propranolol respectively) the increase in heat production has been shown to be due to β_2-adrenoceptor-mediated glycolysis and β_1-adrenoceptor-mediated lipolysis.

During a hypoglycaemic episode the patient appears pale and sweating, but despite the pale skin and increase in catecholamines there is an increase in skin blood flow. This is attributed to the release of vasomotor tone as opposed to active vasodilatation. The sweating is cholinergically mediated rather than adrenergically mediated since hypoglycaemic sweating is still observed in patients after bilateral adrenalectomy.

The heat lost through the increase in peripheral blood flow and sweating exceeds the increase in catecholamine-mediated heat gain and therefore core temperature decreases by around 0.5°C during hypoglycaemia. This fall in core temperature is exaggerated in the cold due to the inhibition of shivering during hypoglycaemia, and in such environmental conditions patients are at severe risk of hypothermia. Thus, the decrease in core temperature during hypoglycaemia is a good example of a disturbance of thermoregulation by an initial non-thermal rather than thermal challenge.

8.1.2 NEUROPATHY

In long-standing cases of diabetes, degenerative changes in the peripheral nerves lead to autonomic dysfunction and disturbances in both sensory and motor function. Autonomic neuropathy can cause impotence in men, bladder dysfunction, diarrhoea, postural hypotension and vomiting. For a recent review of diabetic autonomic neuropathy see Ewing and Clarke (1986).

8.1.3 EXERCISE AND DIABETES

Since physical exercise increases the transport of glucose into cells, in the absence of appreciable amounts of insulin (Chapter 4) exercise reduces the insulin requirements of well-controlled diabetic patients. This effect persists for some time following exercise, and the extent of the effect depends on the intensity of the exercise. However, if the fasting blood glucose exceeds 17–20 mmol $litre^{-1}$, a sign that the diabetes is poorly controlled, moderate exercise lasting 40 min leads to an increase in blood ketones and blood glucose. This seems to be due to an overproduction of glucose. In ketotic diabetics there is also an augmented rise in plasma FFA which is associated with increased muscle use of FFA during exercise. Such an individual may derive 56% of energy from FFA and 6% from ketone metabolism, compared with only 30% from FFA and none from ketone metabolism in well-controlled or non-diabetic individuals during moderate work lasting for 40 min. Thus, the metabolic picture for the ketotic diabetic is one of accelerated metabolism, so that after 40 min of moderate exercise his metabolic profile is rather like that of a well-controlled or non-diabetic individual after 4 h. Thus, exercise is beneficial for the well-controlled patient but it is likely to be detrimental to the ketotic individual.

Plasma catecholamines are higher at rest and during exercise in ketotic compared with well-controlled diabetics. It seems that the poorly controlled diabetic may be more sensitive to sympathoadrenal activity, and this may be related to the more variable metabolic profile of these patients.

In mild, non-insulin-dependent diabetes and in individuals with impaired glucose tolerance, regular endurance exercise, equivalent to running 25–35 km a week, seems to restore glucose tolerance to normal values.

8.1.4 EXERCISE HYPOGLYCAEMIA

The risk of hypoglycaemia during exercise is reduced by using an appropriate site of insulin administration. Exercise increases the absorption of insulin when it is administered into the region of muscle activity. Thus, the injection of insulin into the leg musculature before leg exercise causes a greater rate of insulin disappearance than when insulin is injected into an inactive site such as the abdomen. However, a subcutaneous injection will be less affected by the site of administration than by whether the individual displays thermoregulatory cutaneous vasodilatation.

8.2 Obesity

Obesity is defined as a W/H^2 greater than 30, where W is weight in kg and H is height in metres. This definition is generally valid but may not be appropriate for some athletes (such as shot putters or weight lifters) who may approach this definition of obesity by virtue of an abnormally high body weight in relation to height due to increased muscle mass. Around 7% of people in the UK can be classed as obese by this definition. Whilst there are a number of apparently good reasons why exercise could be beneficial to obese individuals who wish to lose weight, it seems that this is only the case for less serious cases. The following comments on five arguments commonly made to support exercise in the treatment of obesity are based on a review by Garrow (1986).

The first common argument for exercise comes from the observation that dietary restriction leads to a reduction in resting metabolic rate of around 20%. Thus, energy balance is maintained and no weight will be lost by dietary restriction alone. To be in negative energy balance (to lose weight) the individual must increase energy output as well as reduce energy input.

The second commonly made claim is that obese people eat the same amount of energy in food as non-obese people and therefore, if they are not overeating but are overweight, weight loss would occur only through increasing energy expenditure. However, the claim that the energy intakes of obese and non-obese people are the same is not readily substantiated in the literature, not least of all because of inadequate experimental techniques for monitoring energy intake.

A third frequent claim is that inactivity is associated with an increase in food intake and that exercise reduces appetite. There is equivocal evidence to support this claim.

The fourth claim that is often made is that inactivity is associated with a lower lean body mass and therefore a lower resting metabolic rate. However, obese individuals may actually have more lean body mass than non-obese ones, to some extent because of the need for more cytoplasm in the adipose tissue to contain the extra lipid.

Finally, it is claimed that energy expenditure remains elevated for some time after exercise, the so-called excess post-exercise oxygen consumption (EPOC), referred to in Chapter 5. The EPOC may affect energy balance to favour weight loss if the exercise is of a sufficiently high level. However, Garrow (1986) argues that the obese individual has a very low exercise tolerance so that dietary restriction is a more practical solution.

With these points in mind, Garrow (1986) argues that the W/H^2 index should be used as a guide as to whether principally exercise, exercise and dietary restriction or principally dietary restriction should be recommended to an individual who wishes to lose weight (table 8.1).

The question of whether exercise is of benefit for the obese patient remains open for debate and further study. In a recent study, Bahr *et al.* (1987) showed that as little as 20 min of exercise at 70% of $\dot{V}O_2$ max is associated with an EPOC lasting 12 h after finishing exercise. The magnitude of the EPOC is proportional to the duration of the exercise, with the 12 h EPOC being a constant 5% of the

Table 8.1 Recommendations for weight loss according to W/H^2

W/H^2	Recommendation
20–25	Exercise
25–40	Exercise + dietary restriction, with the emphasis on exercise being decreased as the W/H^2 index increases from 25 to 40
40+	Dietary restriction

resting oxygen consumption after 20 min of exercise, 7% after 40 min of exercise and 14% after 80 min of exercise. The 12 h periods in which the EPOC was recorded included two meals, thus suggesting that a prior bout of exercise does not augment the thermic effect of feeding at rest. This is of interest since normal-weight individuals display a higher energy expenditure during exercise in the fed compared with fasted state, a response that is not seen in obese subjects. The value of the 12 h EPOC in the study by Bahr *et al*. was found to be 15% of the total exercise oxygen consumption for all three work periods, each at 70% of maximum oxygen consumption.

One of the likely causes of the EPOC is an effect of circulating catecholamines on increasing the rate of substrate cycles (so-called futile cycles). However, an early study by Svedmyr in 1966 showed that about a third of the thermogenic effect of an infusion of 0.10 $\mu g\ kg^{-1}\ min^{-1}$ adrenaline for 30 min in young volunteers was due to lactate metabolism and a third was due to an increase in fat oxidation. Whereas lactate infusions sufficient to raise the blood lactate by only about 1 mmol $litre^{-1}$ increase oxygen consumption in the absence of a pre-existing increase in plasma catecholamines, the same is not true for fat. The infusion of sufficient triacylglycerol and heparin (the latter to activate the plasma lipoprotein lipase) to raise the blood FFA concentration to the supraphysiological level of 3.0 mmol $litre^{-1}$ has no effect on oxygen consumption in man (Kjekhus *et al*., 1980). However, the concomitant infusion of the sympathomimetic drug isoprenaline in a dose sufficient to raise the heart rate by around 15–20 beats per minute does increase oxygen consumption. None the less, although the increased fat oxidation would be consistent with the decrease in RER during the period of EPOC, it is unlikely to be due to sympathoadrenal activity since plasma noradrenaline and adrenaline rapidly return towards baseline levels and are at sub-thermogenic levels within minutes.

Such a prolonged metabolic response, even after only 20 min of exercise, may be of benefit to the overweight or obese individual. As the following example illustrates, exercise of this magnitude may be of value for such individuals providing there is no increase in energy intake.

The following example describes an overweight individual who presents with a body weight of 84 kg but has an ideal body weight of 70 kg, and in this example it is assumed that the extra weight is adipose tissue. He exercises for 20 min with an oxygen consumption of 1.2 litres min^{-1} (53.6 mmol min^{-1}), corresponding to 70% of his maximum oxygen uptake. Assuming that he uses 100% fat during the exercise (in practice highly unlikely), this results in an increased energy expenditure of 462 kJ day^{-1} (1.2 litres min^{-1} × 19.25 kJ × 20 min). He has a 12 h EPOC which amounts to 15% of the exercising oxygen consumption — equivalent to 3.6 litres [0.15 × (1.2 litres min^{-1} × 20 min)]. The energy cost of this oxygen consumption may be slightly less than during the exercise in view of the greater use of fats, indicated by a reduced RER in recovery. Although this effect is less marked for 20, compared with 40 or 80, minutes of exercise, in this example it is assumed that the individual has switched entirely to fat in recovery. Thus, the EPOC amounts to 69.30 kJ day^{-1} (3.6 litres × 19.25 kJ). The total exercise-induced energy expenditure comprises the energy cost of the exercise plus the EPOC, 531.30 kJ (462 + 69.30 kJ). The physiological energy value of fat is 37.6 kJ g^{-1}, thus the individual is losing 14.1 (531.30 kJ/37.6 kJ g^{-1}) g fat per day. Therefore the loss of 14 kg of fat would take 2 years and 9 months with this daily exercise regime. It would thus seem realistic to advocate only 20 min of exercise per day for the treatment of obesity, providing energy intake (food) is maintained and the duration of the regime is appreciated by the patient.

Interestingly, while there is some evidence that lean individuals automatically increase food intake in proportion to an increase in energy expenditure to maintain body weight, the voluntary energy intake in obese individuals is not changed when habitual activity is increased, thus facilitating the kind of weight loss described above.

8.3 Atherosclerosis and ischaemic heart disease

Over a third of all deaths in the United States of America are attributable to ischaemic heart disease, the most common cause of which is atherosclerosis. The cycle of events leading to atherosclerosis is thought to begin with injury to the arterial endothelial cells which form the practically continuous barrier between the blood and the arterial smooth muscle cells. Initial damage may be caused by high blood pressure or high levels of toxic FFA. This is followed by platelet and fibrin deposition at the site of the injury. In response to stimulation by one or more growth factors from the platelets deposited in the artery wall, there is a proliferation of smooth muscle cells which migrate from the media layer to the region of injury. They thereby cause healing. However, if the individual is a cigarette smoker, has hypertension and/or has hyperlipidaemia, there is likely to be repeated endothelial injury and the development of atherosclerotic plaques which eventually become calcified. Up to 50% of the dry weight of an atherosclerotic plaque is lipid, primarily cholesterol, and the fibrous content is comprised mainly of collagen and elastin. For coronary ischaemia to occur, an obstruction of around 90% of a coronary blood vessel is needed. During exercise, ischaemia will occur with a coronary occlusion of only around 60%. The coronary vessels may also become completely blocked by thrombi formed in the slowly moving blood, and this is especially common in the anterior descending branch of the left coronary artery.

In a myocardial infarction the coronary artery obstruction leads to ischaemia and a regional infarct in which the oxygen-starved cells become necrotic. The dead and weakened cells are unable to function normally. Cardiac contractility is thus compromised and cardiac output is reduced leading to an accumulation of blood in the venous system. Plasma catecholamines increase dramatically during a myocardial infarction, with the rise being proportional to the severity of the infarction. An increase in sympathoadrenal activity is beneficial in terms of its role in maintaining cardiac output and arterial blood pressure at the time of the infarction, but detrimental in the long term since it lowers the threshold for ventricular fibrillation. This may be one reason why β-adrenoceptor antagonists are successful in reducing the risk of further infarction.

8.3.1 PREDISPOSING FACTORS FOR ISCHAEMIC CORONARY HEART DISEASE

The risk of ischaemic coronary heart disease is increased if an individual shows one or more of the known so-called 'risk factors'. The major three risk factors in order of importance are tobacco smoking, high blood cholesterol and high blood pressure. Inactivity is the fourth risk factor.

Smoking

In 1981 it was estimated that 325 000 deaths each year in the USA were due to tobacco smoking. Deaths from cardiovascular disease are more common than from lung cancer in habitual smokers. The occurrence of atherosclerosis in individuals who smoke at least 20 cigarettes a day is three times above that of non-smokers. The severity of the atherosclerosis seems to depend on the number of cigarettes smoked per day and the duration of the smoking habit. There is no clear evidence that either nicotine or carbon monoxide (CO) directly affects the rate of development of atherosclerosis, but both appear to raise blood cholesterol levels. In addition, smokers display abnormal platelet function and impaired fibrinolytic activity, both of which would be expected to increase the rate of atheroma formation.

Nicotine also stimulates the sympathetic nervous system leading to an increase in β_1-adrenoceptor-mediated lipolysis and therefore an increase in circulating FFA. High levels of FFA may produce structural damage to the arterial wall. Further sympathetic responses include cutaneous vasoconstriction, an increase in arterial blood pressure, heart rate and stroke volume (and therefore cardiac output), increased cardiac contractility, myocardial blood flow and myocardial oxygen consumption. A rise in arterial blood pressure and heart rate is seen after smoking only two cigarettes and seems to be due to an increase in sympathetic nervous activity rather than to the humorally mediated effects of the catecholamines. These sympathetic nervous effects are attributed to nicotine rather than CO since non-nicotinic cigarette smoking does not raise the heart rate or blood pressure. For a review of smoking and heart disease see Libow and Schlant (1982).

Lipoprotein metabolism

The lipoproteins account for about 95% of all lipid in the blood, with the rest being in the form of FFA. The lipoproteins are vesicles with a hydrophilic exterior comprising a single layer of phospholipid, cholesterol and around 16 different protein molecules. The interior of the vesicle is hydrophobic and contains cholesterol and triacylglycerol (figure 8.3). The vesicles are termed high or low density, depending on the amounts of lipid (cholesterol and triacylglycerol) relative to protein. The high density lipoproteins (HDL) have a high protein content (around 50%) compared with the low density lipoproteins (LDL) which comprise around 25% protein, and the very low density lipoproteins (VLDL) which comprise only about 10% protein. Correspondingly the lipoprotein density

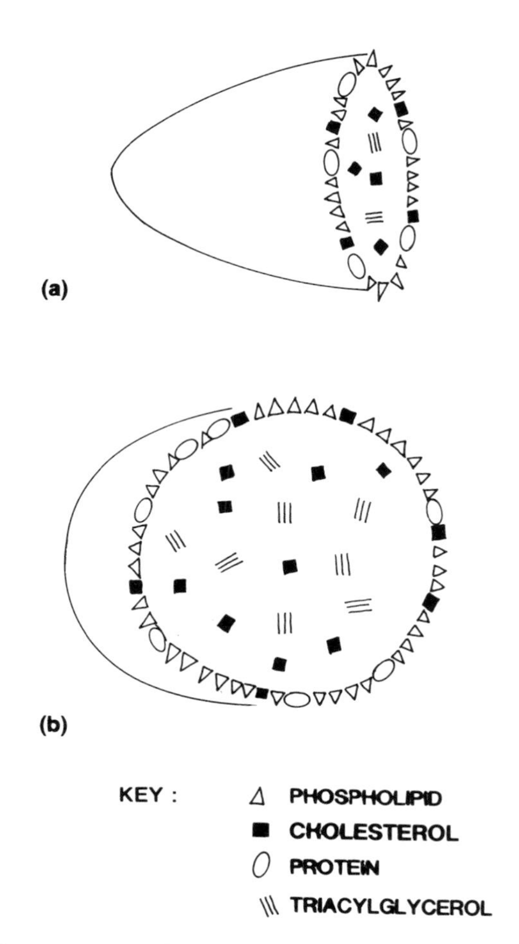

Figure 8.3. Lipoprotein vesicles (schematic). (a) High density lipoprotein; (b) low density lipoprotein.

decreases as the proportion of lipid (which is low in density due to its atomic structure) increases. The HDL vesicles comprise around 20% cholesterol and 5% triacylglycerol compared with 45% cholesterol and 15% triacylglycerol in the LDL vesicles, and 15% cholesterol and 40% triacylglycerol in the VLDL vesicles. The HDL vesicles are smaller and less spherical than the LDL vesicles. The remaining constituent of these vesicles is phospholipid (see table 8.2).

Table 8.2 Percentage composition of lipoprotein vesicles

	HDL	LDL	VLDL
Protein	50	25	10
Cholesterol	20	45	15
Triacylglycerol	5	15	40
Phospholipid	25	15	35

High concentrations of plasma cholesterol, and therefore LDL (which is the major transport vehicle for cholesterol), are associated with an increased risk of coronary heart disease. There is no evidence for a beneficial effect of exercise or training in reducing plasma cholesterol. The LDL changes are variable, with investigations showing a decrease, no change or an increase in LDL concentration through training.

However, individuals engaged in endurance-type exercise at least four times a week and who expend at least 4 MJ in exercise a week, show an increase in HDL which seems to be beneficial in reducing the risk of cardiovascular disease. It seems that of the two subfractions, HDL_2 and HDL_3, only HDL_2 is related (inversely) to the risk of cardiovascular disease, and this is the subfraction increased by regular exercise.

Such exercise training is also found to reduce circulating triacylglycerol levels and the concentration of VLDL in which the major constituent is triacylglycerol. However, such reductions are related to the initial triacylglycerol level, and individuals with high concentrations (> 1.35 mmol litre^{-1}) at the start of training display more dramatic reductions. Such reductions have not been documented in women or children, who generally have low blood triacylglycerol levels.

Hypertension

Hypertension is high arterial blood pressure with the systolic pressure being greater than 20 kPa (150 mm Hg) and diastolic pressure greater than 12 kPa (90 mm Hg). The incidence of hypertension increases with age owing to the age-related rise in arterial blood pressure. Hypertension is the cause of death in around 12% of people in the USA. Theoretically it could be caused by an increase in cardiac output, an increase in peripheral resistance or both (see Chapter 5), but in most cases it is due to an increase in peripheral resistance. The latter may be the result of impaired kidney function leading to increased water retention, and therefore blood volume and blood pressure. A second renal mechanism may be through the production of renin which stimulates the conversion of angiotensin to angiotensin I in the plasma and then to angiotensin II in the lungs. The latter causes vasoconstriction in many vascular beds (including the kidneys) as well as an increase in aldosterone secretion (from the cells of the adrenal cortex). Aldosterone causes an increase in salt and water retention by the kidney. The increased osmolarity draws intracellular fluid into the blood, which (together with increased water retention) causes increased blood pressure.

Most cases of hypertension have no known cause and are classified as 'primary hypertension'. Raised plasma catecholamines have been considered as a possible cause of hypertension as the catecholamines mediate vasoconstriction in many vascular beds. However, the resting plasma catecholamine levels in patients with primary hypertension are similar to those of normotensive controls, although this may apply only to older patients, with other work showing higher circulating catecholamines in plasma from an antecubital vein in young (< 40 years) hypertensive patients. The shortcomings of interpreting measurements of catecholamine concentrations in venous plasma are discussed in Chapter 2. Indeed, direct neuronal recordings of sympathetic muscle nerve activity suggest that while sympathetic activity is modulated by baroreflex mechanisms, thus reflecting acute dynamic changes in diastolic blood pressure, there is no evidence for such a control mechanism in the long-term chronic state of hypertension. Similarly there is no difference in muscle sympathetic nervous activity between hypertensive and control subjects during isometric handgrip exercise. Since the subjects in these last two studies were aged 18–57 years, any effect of age in relation to hypertension and muscle sympathetic activity would have been detected.

There has been much conflicting evidence that regular exercise can effectively reduce arterial blood pressure although the evidence for a reduction in blood pressure is now beginning to outweigh the evidence against. A brief period of post-exercise hypotension may occur immediately following exercise in association with an increase in peripheral blood flow.

Consequences of hypertension The management of high blood pressure is critically important since the increased shearing force on the arterial walls leads to structural damage. This may lead to rupturing of the blood vessel wall which, in cerebral vessels, is the cause of a stroke, and this can be highly debilitating due to loss of cerebral function associated with anoxia in some cells. In addition, hypertension coupled with high blood lipid levels lead to the development of atherosclerosis. The increased work being done by the left ventricle in hypertensive patients leads to hypertrophy of the ventricular muscle cells causing an increase in left ventricular mass. If there is no compensatory increase in vascularization, relative ischaemia will ensue and ultimately the individual may experience angina pectoris. This situation is exacerbated by exercise when extra work is done by the heart since myocardial oxygen consumption can be increased only by changes in blood flow (see Chapter 5).

Prophylactic and therapeutic role of exercise in ischaemic coronary heart disease Regular endurance exercise is thought to be beneficial both for reducing the risk of ischaemic heart disease and in rehabilitation following a myocardial infarction. Evidence for the prophylactic role of exercise in coronary artery disease is largely derived from longitudinal studies which have shown that the incidence of coronary artery disease is higher in sedentary compared with physically active individuals. This association does not necessarily mean that exercise protects an individual against coronary artery disease since it may be that individuals who have subclinical disease are unable to engage in physical activity or prefer not to, or other factors (such as social or cultural) determine job selection, exercise habits and the predisposition to ischaemic coronary disease.

In view of a strong association between occupational activity and the risk of coronary artery disease one might expect to see a reduced risk in individuals who have engaged in sport and athletic activity all their lives. However, interpretation of data from such individuals is likely to be biased as athletes tend to be a self-selected group in terms of genetic make-up and attitudes towards health issues, including nutrition.

There are several mechanisms which could explain why increasing exercise tolerance by physical training is associated with a reduced risk of coronary artery disease. A major factor is probably the effect of training in reducing cardiac work at a given work load, thus reducing myocardial oxygen demand.

Another factor which may reduce the risk of coronary heart disease with training is a decrease in body fat in individuals with a low enough W/H^2 for exercise to be of benefit for weight loss. In this way the energy cost of weight-bearing activities is reduced and myocardial oxygen demands are reduced for a given work load.

Regular exercise is also associated with either retardation or regression of atherosclerosis. This is attributed to:

1. decreased secretion of the atherogenic hormone insulin
2. an increased circulating level of HDL cholesterol
3. increased production of endogenous opiates associated with a reduction in the susceptibility to stress
4. a decrease in arterial blood pressure
5. diminished blood platelet aggregation.

Other arguments for the prophylactic value of exercise in ischaemic heart disease include the development of collateral blood vessels in the heart and widening of the major coronary arteries.

Despite mounting evidence that active individuals are at lower risk of coronary artery disease, at least one study has indicated that individuals who are active in their leisure time are not at a lower risk of cardiovascular disease than inactive people (Wilhelmsen *et al.*, 1976). Those authors therefore argue that to reduce the risk of cardiovascular disease the emphasis should be on stopping smoking and reducing blood lipid by dietary means. Indeed, in order to reduce the risk of sudden death during exercise, those involved in regular vigorous exercise should observe these two recommendations. Although there is some risk of sudden cardiac death during exercise, especially in middle-aged individuals with advanced atherosclerosis, the risk is low.

Physical exercise is now a well accepted part of recovery from myocardial infarction. Typically the exercise programme is started within 2 or 3 days of hospital admission, beginning with passive exercise leading to active low-intensity work. Exercise seems to be beneficial in reducing the risk of reinfarction in cardiac patients due to a decrease in cardiac oxygen requirements for the same amount of muscular work.

References and further reading

Aronow, W.S. (1980) Effect of non-nicotine cigarettes and carbon monoxide on angina. *Circulation*, **61**, 262–265.

Bahr, R., Ingnes, I., Vaage, O., Sejersted, O.M. and Newsholme, E.A. (1987) Effect of duration of exercise on excess postexercise oxygen consumption. *J. Appl. Physiol.*, **62**, 485–490.

Barameyer, J. (1976) Physical activity and coronary collateral development. In *Advances in Cardiology: Physical Activity and Coronary Heart Disease* 18 (eds V. Manninen and P.I. Halonen) S. Karger, Basel, pp. 104–112, 217–230.

Barnard, R.J. (1975) Long term effects of exercise on cardiac function. *Ex. Sports Sci. Rev.*, **3**, 113–133.

Clarke, W.L., Santiago, J.V., Thomas, L., Ben-Galim, E., Haymond, M.W. and Cryer, P.E. (1979) Adrenergic mechanisms in recovering from hypoglycaemia in man: adrenergic blockade. *Am. J. Physiol.*, **236**, E147–E152.

Cryer, P.E., Haymond, A.W., Santiago, J.V. and Shah, S.D. (1976) Norepinephrine and epinephrine release and adrenergic mediation of smoking associated hemodynamic and metabolic events. *N. Engl. J. Med.*, **295**, 573–577.

Cryer, P.E. (1980) Physiology and pathophysiology of the human sympatho-adrenal system. *N. Engl. J. Med.*, **303**, 436–444.

Davidson, N. McD., Corrall, R.J.M., Shaw, T.R.D. and French, E.B. (1977) Observations in man of hypoglycaemia during selective and nonselective beta blockade. *Scott. Med. J.*, **22**, 69–72.

Ewing, D.J. and Clarke, B.F. (1986) Autonomic neuropathy: its diagnosis and prognosis. *Clin. Endocrinol. Metab.*, **15**, 855–888.

Felig, P. and Koivisto, V. (1979) The metabolic response to exercise: implications for diabetes. In *Therapeutics through Exercise* (eds D.T. Lowenthal, K. Bharadwaja and W.W. Oaks), Grune and Stratton, New York, pp. 3–20.

Gale, E.A.M., Bennett, T., Green, J.H. and Macdonald, I.A. (1981) Hypoglycaemia, hypothermia and shivering in man. *Clin. Sci.*, **61**, 463–469.

Garrow, J.S. (1986) Effect of exercise on obesity. *Acta Med. Scand.* Suppl., **711**, 67–73.

Gerich, J., Davis, J., Lorenzi, M., Rizza, R., Bohannon, N., Karam, J., Lewis, S., Kaplan, R., Schultz, T. and Cryer, P. (1979) Hormonal mechanism in recovery from insulin-induced hypoglycaemia. *Am. J. Physiol.*, **236**, E380–E385.

Ginsburg, J. and Paton, A. (1956) Effects in man of insulin hypoglycaemia after adrenalectomy. *J. Physiol.*, **133**, 59P–60P.

Holloszy, J.O., Schultz, J., Kusnierkiewicz, J., Hagberg, J.M. and Ehsani, A.A. (1986) Effects of exercise on glucose tolerance and insulin resistance. *Acta Med. Scand. Suppl.*, **711**, 55–65.

Kannell, W.B. (1981) Update of the role of cigarette smoking in coronary artery disease. *Am. Heart J.*, **101**, 319–328.

Kjekhus, J.K., Ellekjaer, E. and Rinde, P. (1980) The effect of free fatty acids on oxygen consumption in man: the free fatty acid hypothesis. *Scand. J. Lab. Clin. Invest.*, **40**, 63–70.

Koivisto, V.A. and Felig, P. (1978) Effects of leg exercise on insulin absorption in diabetic patients. *N. Engl. J. Med.*, **298**, 79–83.

Libow, M. and Schlant, R.C. (1982) Smoking and heart disease. *Progress in Cardiology* 11 (eds P.N. Yu and J.F. Goodwin), Lea and Febiger, Philadelphia, pp. 131–161.

Link, R.P., Pedersoli, W.M. and Safanie, A.H. (1972) Effect of exercise on development of atherosclerosis in swine. *Atherosclerosis*, **15**, 107–122.

Macdonald, I.A., Bennett, T., Gale, E.A.M., Green, J.H. and Walford, S. (1982) The effect of propranolol or metoprolol on thermoregulation during insulin-induced hypoglycaemia in man. *Clin. Sci.*, **63**, 301–310.

Middleton, W.G. and French, E.B. (1974) Studies of the peripheral vasodilator response to acute insulin-induced hypoglycaemia in man. *Clin. Sci. Mol. Med.*, **47**, 461–470.

Morris, J.N., Kagan, A., Patison, D.C., Gardner, M.J., Raffle, P.A.B. (1966) Incidence and prediction of ischaemic heart disease in London busmen. *Lancet*, **ii**, 553–559.

Nadeau, R.A. and DeChamplain, J. (1979) Plasma catecholamines in acute myocardial infarction. *Am. Heart J.*, **98**, 548–554.

Oberman, A. (1985) Exercise and the primary prevention of cardiovascular disease. *Am. J. Cardiol.*, **55**, 10D–20D.

Paffenbarger Jr, R.S., Hyde, R.T., Hsieh, C-C. and Wing, A.L. (1986) Physical activity, other life-style patterns, cardiovascular disease and longevity. *Acta Med. Scand.* Suppl., **711**, 85–91.

Perper, J.A., Kuller, L.H. and Cooper, M. (1975) Arteriosclerosis of coronary arteries in sudden unexpected deaths. *Circulation*, **51**, Suppl. III, 27–33.

Scheel, K.W., Ingram, L.A. and Wilson, J.L. (1981) Effect of exercise on the coronary and collateral vasculature of beagles with and without coronary occlusion. *Circ. Res.*, **48**, 523–530.

Steiger, J.F. and McCann, D.S. (1982) In vivo platelet aggregation and plasma catecholamines in acute myocardial infarction. *Am. Heart J.*, **104**, 1255–1261.

Sundlöf, G. and Wallin, B.G. (1978) Human muscle nerve sympathetic activity at rest. Relationship to blood pressure and age. *J. Physiol.*, **274**, 621–637.

Svedmyr, N. (1966) Studies on the mechanism for the calorigenic effect of adrenaline in man. *Acta Physiol. Scand.*, **68**, 84–95.

Wahren, J., Felig, P. and Hagenfeldt, L. (1978) Physical exercise and fuel homeostasis in diabetes mellitus. *Diabetologia*, **14**, 213–222.

Wallin, B.G., Mörlin, C. and Hjemdahl, P. (1987) Muscle sympathetic activity and venous plasma noradrenaline concentrations during static exercise in normotensive and hypertensive subjects. *Acta Physiol. Scand.*, **129**, 489–497.

Wilhelmsen, L., Tibblin, G., Aurrell, M., Bjure, J., Ekström-Jodal, B. and Grimby, G. (1976) Physical activity, physical fitness and risk of myocardial infarction. In *Advances in Cardiology: Physical Activity and Coronary Heart Disease* 18 (eds V. Manninen and P.I. Halonen), S. Karger, Basel, pp. 217–230.

Woo, R. and Pi-Sunyer, F.X. (1985) Effect of increased physical activity on voluntary intake in lean women. *Metabolism*, **34**, 836–841.

Woo, R., Garrow, J.S. and Pi-Sunyer, F.X. (1982) Effect of exercise on spontaneous calorie intake in obesity. *Am. J. Clin. Nutr.*, **36**, 370–377.

Index

Acetylcholine *see* Neurotransmission
Action potential 7–8
Adenosinetriphosphate *see* Energy for muscular contraction
Adrenal medulla 42
Adrenaline *see* Catecholamines, circulating
Aerobic metabolism 94–7
Ageing 166–9
  sympathetic cholinergic activity 167–9
  sympathoadrenal activity 166–7
Anaerobic metabolism 91–4
Atherosclerosis 187–95
ATP *see* Energy for muscular contraction
Autonomic nerves *see* Peripheral nervous system
Autonomic neuropathy *see* Diabetes mellitus

Baroreceptors 20, 23, 123
Basal ganglia 67
β-blockers *see* Drugs, β-adrenoceptor antagonism
Blood chemistry 22–3
  pCO_2 22
  pH 22–3
  pO_2 22
Blood flow
  cerebral blood flow 124
  myocardial blood flow 124
  Poiseulle's Law 113
  skeletal muscle blood flow 122–3
  skin blood flow 124
  vasculature 113–16, 122–5, 156–8
  venous return 125
  see also Cardiovascular system, Vascular resistance
Body temperature 142–3
  average body temperature 147, 153
  control mechanisms 153–6
  measurement 150–3
    core temperature 150–2
    skin temperature 152–3
  models of thermoregulation 159–62
    oscillation theory 162
    set point theory 159–60
    variable set point theory 161–2
    see also Sensory receptors, Temperature receptors
Brain *see* Central nervous system
Brainsteam *see* Central nervous system

Carbohydrates 74–5
  glucose 75, 79–85
  liver glycogen 98–100
  muscle glycogen 97–8
  see also Metabolic pathways 79–91
Cardiac output 121–2
Cardiovascular system
  recovery from exercise 125–6
  response to exercise 116–18
    blood flow 122–5
    cardiac output 121–2, 123
    heart rate 118–19
    stroke volume 119–21
  see also Heart, Blood flow 104–26
Catecholamines, circulating 44–7
  blood threshold concentrations 47
  measurement 44–6
Central nervous system 12–24
  basal ganglia 67

brainstem 12
medulla 12, 116, 126
midbrain 13–14
pons 12–13, 126
cerebellum 14, 67
cerebral motor cortex 66
diencephalon 14
epithalamus 15
hypothalamus 14–15
subthalamus 15
thalamus 14
nerve tracts 18
ascending 18–19
descending 23–24, 62–6
spinal cord 17–18
telencephalon 15
limbic system 16, 67
ventricles 16–17
Cerebellum 67
Cerebral motor cortex 66–7
Citric acid cycle *see* Metabolic pathways, TCA cycle
CNS *see* Central nervous system
Control of movement 62–5
central mechanisms 66–8
peripheral mechanisms
final common path 65
Golgi tendon organ 64
motor unit 62
muscle spindle 63–4
reciprocal inhibition 65–6
see also Central nervous system, nerve tracts, descending
Cranial nerves 25

Diabetes mellitus 178–84
exercise 183–84
hypoglycaemia 181–3, 184
neuropathy 183
Diencephalon 14–15
Drugs 172–4
β-adrenoceptor antagonism 173–4

ECG *see* Heart, electrical activity
Efficiency *see* Mechanical efficiency
Electrocardiogram *see* Heart, electrical activity
Electron transport chain, *see* Metabolic pathways, oxidative phosphorylation
Embden–Meyerhof pathway 79–83
Energy for muscular contraction 60–61, 72–4
aerobic metabolism 94–7
anaerobic metabolism 91–4
bond energy 77–8
chemical reactions 78–9
Epithalamus *see* Central nervous system, diencephalon
Extrapyramidal system 24

Fat *see* Free fatty acids, Lipid, Metabolic pathways, β-oxidation of free fatty acids, Obesity
Fatigue *see* Muscle fatigue
FFA *see* Free fatty acids
Fibre types (muscle) 58–60
Fibre types (nerve) 58–60
Free fatty acids 77, 89–91

Gender *see* Sex
Glucagon 101
Gluconeogenesis 84–5
Glucose 75, 79–85
see also Metabolic pathways
Glycogen 97–100
see also Metabolic pathways
Glycolysis 79–83

Haemoglobin 128–31
Heart 104–112
autonomic innervation 107–8
Bainbridge reflex 112
cardiac muscle 106–7
electrical activity 108–112
electrocardiogram 109–112
sinus arrhythmia 112
sounds 106
stretch reflex 112
structure and function 104–6
Heart rate 118–9
Heat balance 143–9
heat balance equation 146–9

Heat balance (*cont'd*)
heat exchange 143–4
heat gain 143
heat loss 145–6
Hypertension 191–3
Hypoglycaemia 181–3, 184
Hypothalamus 14–15, 30, 154–5

Insulin 101
see also Diabetes mellitus
Ischaemic heart disease 187–95
predisposing factors 188–95
hypertension 191–3
lipoproteins 189–91
smoking 188–9
prophylactic and therepeutic role of exercise 193–5

Ketone bodies 179
Kinaesthesia *see* Sensory receptors, position sense
Krebs' cycle *see* Metabolic pathways, TCA cycle

Limbic system 16, 67
Lipid *see* Free fatty acids, Ischaemic heart disease, predisposing factors, lipoproteins, Metabolic pathways, β-oxidation of free fatty acids 75, 100–101, 189–91
Lipoproteins 189–91

Mechanical efficiency 61–2
Medulla 12, 116, 126
Metabolic pathways 79–91
β-oxidation of free fatty acids 89–91
gluconeogenesis 84–5
glycolysis 79–83
oxidative phosphorylation 88–9
TCA cycle 85–9
Midbrain 13–14
Motor control *see* Control of movement
Motor unit 62
Muscle *see* Heart, cardiac muscle, Vascular resistance, smooth muscle, Skeletal muscle

Myoglobin 133

Nervous system
nerve fibre classification 11
organization 10–11
see also Central nervous system, Peripheral nervous system 1–30
Neurone 1–3
axon 3
cell body 1–2
nerve ending 3
Neurotransmission
acetylcholone 49–52
noradrenaline 34–42
purinergic nerves 53
see also Synapse 32–4
Noradrenaline *see* Catecholamines, circulating, Neurotransmission

Obesity 184–7
β-oxidation of free fatty acids 89–91
Oxidative phosphorylation 88–9
Oxygen consumption
cardiac muscle 135–6
skeletal muscle 135
whole body, exercise 137–9
see also Respiration 134–9
Oxygen dissociation curve 129–31

pCO_2 *see* Blood chemistry
Peripheral nervous system 25–30
autonomic nerves 27–30, 68–71
parasympathetic autonomic nervous system 28–30
sympathetic autonomic nervous system 28, 29
cranial nerves 25
spinal nerves 25–26
pH *see* Blood chemistry
pO_2 *see* Blood chemistry
Poiseulle's Law 113
Pons 12–13, 126
Proprioception *see* Sensory receptors, position sense
Purinergic nerves 53
Pyramidal system 24

Reflex arc, autonomic 27–28
Reflex arc, somatic 26–7, 63
Respiration
pulmonary ventilation 126–8
ventilatory threshold 128
tissue respiration 131–4
carbon dioxide transport, blood 133–4
oxygen delivery 132
oxygen extraction 132
oxygen transport, blood 128–31
oxygen transport, muscle 132–3
see also Oxygen consumption 126–39
Risk factors *see* Ischaemic heart disease, predisposing factors

Sensory receptors
baroreceptors 20, 23, 123
position sense 20
temperature receptors 20
see also Blood chemistry 19–23
Sex 169–70
sympathetic cholinergic activity 169–70
sympathoadrenal activity 169
Shivering 159
Skeletal muscle 56–62
structure and function 56–8
types of fibre 58–60
Smoking 188–9
Smooth muscle 115
Spinal cord 17–18
Spinal nerves 25–6
Stroke volume 119–21
Subthalamus 15
Sweating 145, 148–9, 158
Sympathetic autonomic nervous system, assessment 42
Sympathoadrenal activity 42–9
Synapse 3–7
chemical synapse 3
electrical synapse 3
excitatory synapse 6
final common path 65
inhibitory synapse 6
neurotransmitters 4, 32

TCA cycle 85–9
Telencephalon 15–17
Thalamus 14
Training 138, 170–2, 174, 194
sympathetic cholinergic activity 172
sympathodrenal activity 170–2
Tricarboxylic acid cycle *see* Metabolic pathways, TCA cycle

Vascular resistance smooth muscle 115
Vasculature 114–116
autonomic innervation 116–117
Venous return 125
Ventricles (brain) *see* Central nervous system
Ventricles (heart) *see* Heart, structure and function
VO_2 *see* Oxygen consumption

The Autonomic Nervous System and Exercise